Applied Asymptotics: Case Studies in Small-Sample Statistics

In fields such as biology, medical sciences, sociology and economics researchers often face the situation where the number of available observations, or the amount of available information, is sufficiently small that approximations based on the normal distribution may be unreliable. Theoretical work over the last quarter-century has led to new likelihood-based methods that yield very accurate approximations in finite samples, but this work has had limited impact on statistical practice. This book illustrates by means of realistic examples and case studies how to use the new theory, and investigates how and when it makes a difference to the resulting inference. The treatment is oriented towards practice and is accompanied by code in the R language which enables the methods to be applied in a range of situations of interest to practitioners. The analysis includes some comparisons of higher order likelihood inference with bootstrap and Bayesian methods.

ALESSANDRA BRAZZALE is a Professor of Statistics at the Università degli Studi di Modena e Reggio Emilia.

ANTHONY DAVISON is a Professor of Statistics at the Ecole Polytechnique Fédérale de Lausanne.

NANCY REID is a University Professor of Statistics at the University of Toronto.

Already published

Applied Asymptotics
Case Studies in Small-Sample
Statistics

A. R. Brazzale
Università degli Studi di Modena e Reggio Emilia

A. C. Davison
Ecole Polytechnique Fédérale de Lausanne

N. Reid
University of Toronto

CAMBRIDGE
UNIVERSITY PRESS

CAMBRIDGE
UNIVERSITY PRESS

Shaftesbury Road, Cambridge CB2 8EA, United Kingdom

One Liberty Plaza, 20th Floor, New York, NY 10006, USA

477 Williamstown Road, Port Melbourne, VIC 3207, Australia

314–321, 3rd Floor, Plot 3, Splendor Forum, Jasola District Centre, New Delhi – 110025, India

103 Penang Road, #05–06/07, Visioncrest Commercial, Singapore 238467

Cambridge University Press is part of Cambridge University Press & Assessment,
a department of the University of Cambridge.

We share the University's mission to contribute to society through the pursuit of
education, learning and research at the highest international levels of excellence.

www.cambridge.org
Information on this title: www.cambridge.org/9780521847032

First published 2007

A catalogue record for this publication is available from the British Library

ISBN   978-0-521-84703-2   Hardback

Contents

Preface

The likelihood function plays a central role in both statistical theory and practice. Basic results about likelihood inference, which we call first order asymptotics, were developed in fundamental work by R. A. Fisher during the 1920s, and now form an essential and widely taught part of both elementary and advanced courses in statistics. It is less well known that Fisher later proposed a more refined approach, which has been developed over the past three decades into a theory of higher order asymptotics. While this theory leads to some extremely accurate methods for parametric inference, accounts of the theory can appear forbidding, and the results may be thought to have little importance for statistical practice.

The purpose of this book is dispel this view, showing how higher order asymptotics may be applied in realistic examples with very little more effort than is needed for first order procedures, and to compare the resulting improved inferences with those from other approaches. To do this we have collected a range of examples and case studies, provided details on the implementation of higher order approximations, and compared the resulting inference to that based on other methods; usually first order likelihood theory, but where appropriate also methods based on simulation. Our examples are nearly all derived from regression models for discrete or continuous data, but range quite widely over the types of models and inference problems where likelihood methods are applied.

In order to make higher order methods accessible, we have striven for as simple an exposition as we thought feasible, aiming for heuristic explanation rather than full mathematical rigour. We do not presuppose previous knowledge of higher order asymptotics, key aspects of which are explained early in the book. The reader is assumed to have knowledge of basic statistics including some central classes of models, and some experience of standard likelihood methods in practice. We intend that the book be useful for students of statistics, practising statisticians, and data analysts, as well as researchers interested in a more applied account of the methods than has so far been available. Our effort has been made practicable by software developed by Alessandra Brazzale and Ruggero Bellio over many years, of which the hoa package bundle now available in R is the culmination. This software is extensively used throughout the book, and the ideas behind the hoa packages, described in Chapter 9, formed the basis for our approaches to programming when new software was needed for some of the examples. The `hoa` package bundle and other materials may be obtained from the book's web page

`http://statwww.epfl.ch/AA`

This book could not have been written without the help of many colleagues. We thank particularly Sven Bacher, Douglas Bates, Ruggero Bellio, Nathalie Chèvre, David Cox, Don Fraser, Torsten Hothorn, Alessandra Salvan, Ana-Maria Staicu, Charlotte Vandenberghe, and members of the R Development Core Team, for access to data, fruitful collaboration, helpful discussion, valuable comments on the material, and help with computational aspects of our work. David Tranah and Diana Gillooly of Cambridge University Press have been supportive and patient editors. We thank also the following institutions for financial and material support: the EPFL; the University of Padova; the University of Toronto; the Italian National Research Council, and in particular its Institute of Biomedical Engineering, where much of Alessandra Brazzale's work was performed; the Italian Ministry of Education, University and Research; the Canadian Natural Sciences and Engineering Research Council; and the Swiss National Science Foundation. We thank also our friends and families for their enduring encouragement and support.

A. R. Brazzale, A. C. Davison and N. Reid
Reggio Emilia, Lausanne and Toronto

1

Introduction

This book is about the statistical analysis of data, and in particular approximations based on the likelihood function. We emphasize procedures that have been developed using the theory of higher order asymptotic analysis and which provide more precise inferences than are provided by standard theory. Our goal is to illustrate their use in a range of applications that are close to many that arise in practice. We generally restrict attention to parametric models, although extensions of the key ideas to semi-parametric and non-parametric models exist in the literature and are briefly mentioned in contexts where they may be appropriate. Most of our examples consist of a set of independent observations, each of which consists of a univariate response and a number of explanatory variables.

Much application of likelihood inference relies on *first order asymptotics*, by which we mean the application of the central limit theorem to conclude that the statistics of interest are approximately normally distributed, with mean and variance consistently estimable from the data. There has, however, been great progress over the past twenty-five years or so in the theory of likelihood inference, and two main themes have emerged. The first is that very accurate approximations to the distributions of statistics such as the maximum likelihood estimator are relatively easily derived using techniques adapted from the theory of asymptotic expansions. The second is that even in situations where first order asymptotics is to be used, it is often helpful to use procedures suggested by these more accurate approximations, as they provide modifications to naive approaches that result in more precise inferences. We refer throughout to these two developments as *higher order asymptotics*, although strictly speaking we mean 'higher order asymptotic theory applied to likelihood inference' – of course there are many developments of higher order asymptotics in other mathematical contexts.

Asymptotic theory in statistics refers to the limiting distribution of a summary statistic as the amount of information in the data increases without limit. In the simplest situations this means that the sample size increases to infinity, and in more complex models entails a related notion of accumulation of information, often through independent replications of an assumed model. Such asymptotic theory is an essential part of statistical methodology, as exact distributions of quantities of interest are rarely available. It serves in the first instance to check if a proposed inferential method is sensible, that is, it provides what would be regarded as correct answers if there were an unlimited amount of data related to the problem under study. Beyond this minimal requirement asymptotic theory serves to provide approximate answers when exact ones are unavailable. Any achieved amount of information is of course finite, and sometimes quite small, so these approximations need to be checked against exact answers when possible, to be verified by simulation, and to stand the test of practical experience.

The form of the limiting distribution of a statistic is very often obtained from the first term in an asymptotic expansion, higher order terms decreasing to zero as the sample size or amount of information becomes infinite. By considering further terms in the expansion, we may hope to derive approximations that are more accurate for cases of fixed sample size or information. An example of this is the analysis of the distribution of the average of a sample of independent, identically distributed random variables with finite mean and variance. Under some conditions, an asymptotic expansion of the moment generating function of the average, suitably standardized, has as leading term the moment generating function of a standard normal random variable. Incorporating higher order terms in the expansion directly leads to the Edgeworth approximation, and indirectly to the saddlepoint approximation; both of these underlie the theory used in this book. Asymptotic expansions are not convergent series, so including further terms in an expansion does not guarantee a more accurate approximation. In the absence of any uniform error bounds on the omitted terms, detailed examination of examples is needed. Among several asymptotically equivalent approximations, one will sometimes emerge as preferable, perhaps on the basis of detailed numerical work, or perhaps from more general arguments.

One intriguing feature of the theory of higher order likelihood asymptotics is that relatively simple and familiar quantities play a central role. This is most transparent in the tail area approximations for a single scalar parameter, where a combination of the likelihood ratio statistic and the score or Wald statistic leads to remarkably more accurate results than those available from first order theory, but is also the case in more complex models. In a very general way, the likelihood function emerges as an approximately pivotal quantity, that is, a function of the data and the parameter of interest that has a known distribution. In this sense the approximations can be viewed as generalizing Fisher's (1934) result for location models.

In nearly all applications of statistical inference the models used are provisional, in the sense that they are not derived from a precise and widely-accepted scientific theory, although the most useful models will be broadly consistent with theoretical predictions. Both the usual first order likelihood inferences and the type of higher order approximations we shall discuss depend for their validity on the assumed correctness of the model. The point is often made, quite reasonably, that the gains from improved approximations are potentially outweighed by sensitivity to the modelling assumptions. This will have more or less force in particular examples, depending on the model and the application. For example, linear regression has proved to be a useful starting point in innumerable situations, without any notion of a linear model being the 'true' underlying mechanism, so it is worthwhile to ensure the accuracy of the inferential procedure, insofar as this is possible. Furthermore, the availability of easily computed higher order approximations can enable an investigation of the stability of the conclusions to a variety of models. However if it is likely that data may contain one or more extreme outliers, then understanding the origin and influence of these will usually be of more practical importance than highly accurate calculation of confidence limits in a model that ignores outliers.

Most of the higher order approximations we discuss in the examples are for P-values; that is, the probability of observing a result as or more extreme than that observed, under an assumed value for the parameter of interest. The P-value function provides confidence limits at any desired level of confidence. In problems with discrete data the issue of continuity correction arises, and this will be discussed as needed, particularly in Chapter 4.

In Chapter 2 we give a very brief introduction to the main approximations used in the examples. We review first order results, and describe the construction of first and higher order approximations to P-values and posterior probabilities.

The heart of the book is a collection of examples, or case studies, illustrating the use of higher order asymptotics. We have tried to choose examples that are simple enough to be described briefly, but complex enough to be suggestive for applied work more generally. In Chapter 3 we present some elementary one-parameter models, chosen to illustrate the potential accuracy of the procedures in cases where one can perform the calculations easily, and some one- and two-sample problems. Chapter 4 illustrates regression with categorical responses; in particular logistic regression, several versions of 2×2 tables, and Poisson regression. Chapter 5 illustrates linear and non-normal linear regression, nonlinear regression with normal errors, and nonlinear regression with non-constant variance. In Chapters 3, 4 and 5 we try to emphasize the data as much as possible, and let the models and analysis follow, but the data sets are generally chosen to illustrate particular methods. Chapter 6 treats in depth analysis of data that arose in collaboration with colleagues. In these examples the emphasis is on model building and inference; while higher order approximations are used they are not the main focus of the discussion. Chapter 7 takes a different approach: in order to illustrate the wide range of examples that can be treated with higher order asymptotics, we discuss a number of model classes, and use the data simply to illustrate the calculations.

A more detailed discussion of the theoretical aspects of higher order approximation is given in Chapter 8. This topic has a large literature, and some of it is rather formidable. Issues that must be faced almost immediately are the role of conditioning in inference, and definitions of sufficiency and ancillarity in the presence of nuisance parameters. It is also necessary to consider the likelihood function as a function of both the parameter and the data, a viewpoint that is usually unfamiliar to those outside the area. Our goal is to make this literature somewhat more accessible for use in applications of reasonable sophistication, both to assess to what extent this is useful, and to provide some guidance for those seeking to apply these methods in related problems. For the derivations of the results, we direct the reader to a number of books and review papers in the bibliographic notes.

In Chapter 9 we provide some details of our numerical work, which is based on the R package bundle hoa originally developed for S-PLUS and described by Brazzale (2000). There are a number of general points relevant to any implementation, and some slightly more specialized issues in this approach. For some examples, we needed additional code, and we have provided this on the book's web page (see Preface) where we thought it may be useful.

Chapter 10 contains a variety of problems and further results based on the material, and an appendix sketches the asymptotic methods that form the basis for the development of higher order asymptotic inference.

Bibliographic notes

We have been strongly motivated in this effort by the book *Applied Statistics* by Cox and Snell (1981), whose introductory chapters provide an excellent introduction to the role of statistical models and theory in applied work. See also the introductory chapter of Cox and Wermuth (1996), and Davison (2003, Chapter 12). The role of asymptotics in the theory of statistics is surveyed by Reid (2003) and Skovgaard (2001).

We give more thorough notes related to the literature on higher order asymptotics at the end of Chapter 8, but the main book-length treatments of likelihood-based asymptotics are the monographs by Barndorff-Nielsen and Cox (1989, 1994), Pace and Salvan (1997) and Severini (2000a). An overview is given in Brazzale (2000). The hoa package bundle is described in Brazzale (2005) and the computing strategy for higher order approximations is discussed in Brazzale (1999) and Bellio and Brazzale (2003).

2

Uncertainty and approximation

2.1 Introduction

In the examples in later chapters we use parametric models almost exclusively. These models are used to incorporate a key element of statistical thinking: the explicit recognition of uncertainty. In frequentist settings imprecise knowledge about the value of a single parameter is typically expressed through a collection of confidence intervals, or equivalently by computation of the P-values associated with a set of hypotheses. If prior information is available then Bayes' theorem can be employed to perform posterior inference.

In almost every realistic setting, uncertainty is gauged using approximations, the most common of which rely on the application of the central limit theorem to quantities derived from the likelihood function. Not only does likelihood provide a powerful and very general framework for inference, but the resulting statements have many desirable properties.

In this chapter we provide a brief overview of the main approximations for likelihood inference. We present both first order and higher order approximations; first order approximations are derived from limiting distributions, and higher order approximations are derived from further analysis of the limiting process. A minimal amount of theory is given to structure the discussion of the examples in Chapters 3 to 7; more detailed discussion of asymptotic theory is given in Chapter 8.

2.2 Scalar parameter

In the simplest situation, observations $y_1, \ldots, y_n$ are treated as a realization of independent identically distributed random variables $Y_1, \ldots, Y_n$ whose probability density function $f(y; \theta)$ depends on an unknown scalar parameter θ. Let $\ell(\theta) = \sum \log f(y_i; \theta)$ denote the log likelihood based on the observations, $\widehat{\theta}$ the maximum likelihood estimator, and $j(\theta) = -\partial^2 \ell(\theta)/\partial \theta^2$ the observed information function. Below we adopt the convention that additive constants that do not depend on the parameter may be neglected when log likelihoods are defined, and we suppress them without further comment.

Likelihood inference for θ is typically based on the

$$\text{likelihood root,} \quad r(\theta) = \text{sign}(\widehat{\theta} - \theta)\left[2\left\{\ell(\widehat{\theta}) - \ell(\theta)\right\}\right]^{1/2}; \tag{2.1}$$

$$\text{score statistic,} \quad s(\theta) = j(\widehat{\theta})^{-1/2}\partial\ell(\theta)/\partial\theta; \quad \text{or} \tag{2.2}$$

$$\text{Wald statistic,} \quad t(\theta) = j(\widehat{\theta})^{1/2}(\widehat{\theta} - \theta). \tag{2.3}$$

These quantities are functions of the data and the parameter, so strictly speaking should not be called *statistics* unless θ is fixed at a particular value. Note the role of the observed Fisher information, $j(\widehat{\theta})$, in calibrating the distance of $\widehat{\theta}$ from the true value θ, and the distance of the score function $\ell_\theta(\theta) = \partial\ell(\theta)/\partial\theta$ from its expectation of zero.

Under suitable conditions on the parametric model and in the limit as $n \to \infty$, each of these statistics has, under $f(y; \theta)$, an asymptotic standard normal, $N(0, 1)$, distribution. A closely related quantity, the likelihood ratio statistic

$$w(\theta) = r(\theta)^2 = 2\left\{\ell(\widehat{\theta}) - \ell(\theta)\right\}, \tag{2.4}$$

has an asymptotic chi-squared distribution with one degree of freedom, χ_1^2. Sometimes the log likelihood is multimodal, so it is useful to graph it or equivalently $w(\theta)$. In most cases, however, $\ell(\theta)$ has a single prominent mode around $\widehat{\theta}$, and then $r(\theta)$, $s(\theta)$ and $t(\theta)$ are decreasing functions of θ in the region of that mode. Various alternative forms of $s(\theta)$ and $t(\theta)$ may be defined by replacing $j(\widehat{\theta})$ by $j(\theta)$, $i(\widehat{\theta})$ or $i(\theta)$, where $i(\theta) = \text{E}\{j(\theta)\}$ is the expected information; in fact the name 'Wald statistic' used in (2.3) is a slight misnomer as the version standardized with $i(\theta)^{1/2}$ is more correctly associated with Abraham Wald. Under suitable conditions similar distributional results apply far beyond independent identically distributed observations, and for vector parameters – in particular, $w(\theta)$ has an asymptotic χ_d^2 distribution when θ is of fixed dimension d. For now, however, we continue to suppose that θ is scalar.

An important variant of the likelihood root is the *modified likelihood root*

$$r^*(\theta) = r(\theta) + \frac{1}{r(\theta)}\log\left\{\frac{q(\theta)}{r(\theta)}\right\}, \tag{2.5}$$

which is based on so-called *higher order asymptotics* that are described in Chapter 8. The modified likelihood root combines the likelihood root with the score statistic, if $q(\theta) = s(\theta)$, with the Wald statistic, if $q(\theta) = t(\theta)$, or with variants of these which depend on the context. We shall see below that normal approximation to the distribution of $r^*(\theta)$ can provide almost exact inferences for θ, when these are available.

Any of (2.1)–(2.5) or their variants may be used to set confidence intervals or compute a P-value for θ. Suppose, for instance, that it is desired to use the Wald statistic (2.3) to test the null hypothesis that θ equals some specified θ_0 against the alternative $\theta > \theta_0$. As a large positive value of $t(\theta_0)$ relative to the $N(0, 1)$ distribution will give evidence against the null hypothesis, the corresponding P-value is $1 - \Phi\{t(\theta_0)\}$, where Φ denotes the standard normal distribution function. Likewise a confidence interval may be constructed using those values of θ most consistent with this distribution; for example, the limits of an equi-tailed $(1 - 2\alpha)$ interval $(\theta_\alpha, \theta^\alpha)$ for θ_0 are given by

$$\Phi\{t(\theta_\alpha)\} = 1 - \alpha, \quad \Phi\{t(\theta^\alpha)\} = \alpha,$$

or equivalently

$$t(\theta_\alpha) = z_{1-\alpha}, \quad t(\theta^\alpha) = z_\alpha,$$

where z_α is the α quantile of the standard normal distribution. The resulting interval,

$$(\theta_\alpha, \theta^\alpha) = (\widehat{\theta} - z_{1-\alpha} j(\widehat{\theta})^{-1/2}, \widehat{\theta} - z_\alpha j(\widehat{\theta})^{-1/2}),$$

is highly convenient and widely used because it can be computed for any desired α using only $\widehat{\theta}$ and $j(\widehat{\theta})^{-1/2}$. Often in practice a 95% confidence interval is sought; then $\alpha = 0.025$, $z_\alpha = -1.96$, and the interval has limits of the familiar form $\widehat{\theta} \pm 1.96 j(\widehat{\theta})^{-1/2}$.

Similar computations yield intervals based on $r(\theta)$, $s(\theta)$ and $r^*(\theta)$, though typically numerical solution is required. Suppose, for example, that a $(1 - 2\alpha)$ confidence interval is to be based on $r^*(\theta)$. Then one possibility is to compute the values of $r^*(\theta)$ for a grid of values of θ, to fit a spline or other interpolating curve to the resulting pairs $(r^*(\theta), \theta)$, and to read off those values of θ corresponding to $r^*(\theta) = z_\alpha, z_{1-\alpha}$. We often plot such curves or the corresponding probabilities $\Phi\{r^*(\theta)\}$, and call the plots *profiles*. An alternative to $\Phi\{r^*(\theta)\}$ which has the same asymptotic properties is the *Lugannani–Rice formula*

$$\Phi\{r(\theta)\} + \left\{ \frac{1}{r(\theta)} - \frac{1}{q(\theta)} \right\} \phi\{r(\theta)\}, \tag{2.6}$$

where ϕ denotes the standard normal density function.

When the likelihood is unimodal, confidence intervals based on $r(\theta)$ and on $w(\theta)$ are the same, and equal to

$$\{\theta : w(\theta) \leq c_{1,1-2\alpha}\}. \tag{2.7}$$

Here and below we use $c_{\nu,1-2\alpha}$ to denote the $(1 - 2\alpha)$ quantile of the χ_ν^2 distribution. If $\ell(\theta)$ is multimodal then (2.7) may be a union of disjoint intervals. If the random variable Y is continuous then the chi-squared approximation to the distribution of $w(\theta)$ is improved if a *Bartlett adjustment* is used: $w(\theta)$ is replaced in (2.7) by $w(\theta)/(1 + b/n)$, where the *Bartlett correction* b is computed from $E\{w(\theta)\}$.

A quantity which depends both on the data and on the parameter, and whose distribution is known, is called a *pivot*, or *pivotal quantity*. The previous paragraphs describe the use of the quantities $r(\theta)$, $s(\theta)$, $t(\theta)$, $r^*(\theta)$ and $w(\theta)$, regarded as functions of the data and of θ, as *approximate* pivots; their asymptotic distributions are used to set confidence intervals or compute P-values for inference on θ. For brevity we will refer to $r(\theta)$, $s(\theta)$ and so forth as pivots even when their exact distributions are unknown. We call the functions of θ obtained by computing the P-value for a range of values of θ *significance functions*; an example is $\Phi\{r(\theta)\}$. As we now illustrate, the usefulness of these significance functions will depend on the accuracy of the underlying distributional approximations.

Illustration: Exponential data

Suppose that a sample $y_1, \ldots, y_n$ is available from the exponential density

$$f(y; \theta) = \theta \exp(-\theta y), \quad y > 0, \theta > 0,$$

and that a 95% confidence interval is required for θ. The log likelihood is

$$\ell(\theta) = n(\log \theta - \theta \bar{y}), \quad \theta > 0,$$

where $\bar{y} = (y_1 + \cdots + y_n)/n$ is the sample average. Here $\ell(\theta)$ is unimodal with maximum at $\hat{\theta} = 1/\bar{y}$ and observed information function $j(\theta) = n/\theta^2$, and

$$
\begin{aligned}
r(\theta) &= \operatorname{sign}(1 - \theta \bar{y}) \left[2n \{\theta \bar{y} - \log(\theta \bar{y}) - 1\}\right]^{1/2}, \\
s(\theta) &= n^{1/2} \{1/(\theta \bar{y}) - 1\}, \\
t(\theta) &= n^{1/2}(1 - \theta \bar{y}).
\end{aligned}
$$

In this exponential family model it is appropriate to take $q(\theta) = t(\theta)$ in the construction of $r^*(\theta)$. The Bartlett correction is readily obtained on noting that $\mathrm{E}\{w(\theta)\} = 2n\{\log n - \Psi(n)\}$, where $\Psi(n) = \mathrm{d} \log \Gamma(n)/\mathrm{d} n$ is the digamma function.

The quality of the approximations outlined above may be assessed using the exact pivot $\theta \sum Y_i$, whose distribution is gamma with unit scale and shape parameter n.

Consider an exponential sample with $n = 1$ and $\bar{y} = 1$; then $j(\hat{\theta}) = 1$. The log likelihood $\ell(\theta)$, shown in the left-hand panel of Figure 2.1, is unimodal but strikingly asymmetric, suggesting that confidence intervals based on an approximating normal distribution for $\hat{\theta}$ will be poor. The right-hand panel is a chi-squared probability plot in which the ordered values of simulated $w(\theta)$ are graphed against quantiles of the χ_1^2 distribution – if the simulations lay along the diagonal line $x = y$, then this distribution would be a perfect fit. The simulations do follow a straight line rather closely, but with slope $(1 + b/n)$, where $b = 0.1544$. This indicates that the distribution of the Bartlett-adjusted likelihood ratio statistic $w(\theta)/(1 + b/n)$ would be essentially χ_1^2. The 95% confidence intervals for θ based on the unadjusted and adjusted likelihood ratio statistics are $(0.058, 4.403)$ and $(0.042, 4.782)$ respectively.

The left-hand panel of Figure 2.2 shows the pivots $r(\theta)$, $s(\theta)$, $t(\theta)$, and $r^*(\theta)$. The limits of an equi-tailed 95% confidence interval for θ based on $r(\theta)$ are given by the intersections of the horizontal lines at ± 1.96 with the curve $r(\theta)$, yielding $(0.058, 4.403)$; this equals the interval given by $w(\theta)$, as of course it should. The 95% interval based on the exact pivot $\theta \sum y_i$ is $(0.025, 3.689)$, while the interval based on $t(\theta)$ is

$$\hat{\theta} \pm 1.96 j(\hat{\theta})^{-1/2} = 1 \pm 1.96 = (-0.96, 2.96),$$

which is a very unsatisfactory statement of uncertainty for a positive quantity. The 95% confidence interval based on $r^*(\theta)$ is $(0.024, 3.705)$, very close to the exact one. This illustrates one important advantage of $r(\theta)$ and $r^*(\theta)$; intervals based on these

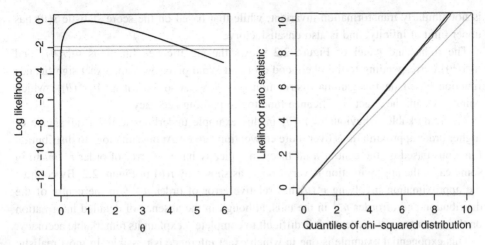

Figure 2.1 Likelihood inference for exponential sample of size $n = 1$. Left: log likelihood $\ell(\theta)$. Intersection of the function with the two horizontal lines gives two 95% confidence intervals for θ: the upper line is based on the χ_1^2 approximation to the distribution of $w(\theta)$, and the lower line is based on the Bartlett-adjusted statistic. Right: comparison of simulated values of likelihood ratio statistic $w(\theta)$ with χ_1^2 quantiles. The χ_1^2 approximation is shown by the line of unit slope, while the $(1 + b/n)\chi_1^2$ approximation is shown by the upper straight line.

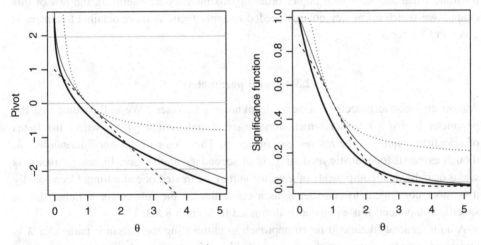

Figure 2.2 Approximate pivots and P-values based on an exponential sample of size $n = 1$. Left: likelihood root $r(\theta)$ (solid), score pivot $s(\theta)$ (dots), Wald pivot $t(\theta)$ (dashes), modified likelihood root $r^*(\theta)$ (heavy), and exact pivot $\theta \sum y_i$ indistinguishable from the modified likelihood root. The horizontal lines are at $0, \pm 1.96$. Right: corresponding significance functions, with horizontal lines at 0.025 and 0.975.

must lie within the parameter space. A further advantage is the invariance of these intervals to parameter transformation: for example, the likelihood root for θ^{-1} yields the 95% confidence interval $(1/4.403, 1/0.057)$, which is obtained simply by applying the reciprocal transformation to the likelihood root interval for θ. The interval based on $t(\theta)$

is not similarly transformation-invariant, while that based on the score statistic $s(\theta)$ has upper limit at infinity, and is also unsatisfactory.

The right-hand panel of Figure 2.2 shows the significance functions $\Phi\{r(\theta)\}$ and $\Phi\{t(\theta)\}$ corresponding to the likelihood root and Wald pivot, with the exact significance function based on the gamma distribution of $\theta \sum y_i$. Also shown is $\Phi\{r^*(\theta)\}$, which coincides with the exact significance function to plotting accuracy.

The remarkable behaviour of $r^*(\theta)$ in this example underlines a major advantage of higher order approximations over more conventional ones. Approximations to distribution functions based on the central limit theorem typically have an error of order $n^{-1/2}$, so in some cases the approximation is very poor – as shown by $t(\theta)$ in Figure 2.2. By contrast the approximation involving $r^*(\theta)$ has relative error of order $n^{-3/2}$ in the centre of the distribution and of order n^{-1} in the tails, although in the absence of detailed information on the constants in these terms it is difficult to completely explain its remarkable accuracy.

This exponential example is one in which exact inference is possible. In most realistic cases approximation based on asymptotic arguments is required, however, and then the quality of the inference will depend on the accuracy of the approximation. As we have seen above, the same ingredients can be combined in different ways to give markedly different results – in particular, inference based on $r^*(\theta)$ appeared to recover the exact results almost perfectly. The applications in later chapters of this book are intended to illustrate when and how such higher order approximations are useful. In the rest of this chapter we sketch a few key notions needed to apply them. A more detailed treatment is given in Chapter 8.

2.3 Several parameters

Almost all realistic models have several unknown parameters. We will assume that the parameter is a $d \times 1$ vector which may be expressed as $\theta = (\psi, \lambda)$, where the focus of scientific enquiry is the *interest parameter* ψ. The vector of *nuisance parameters* λ, though essential for realistic modelling, is of secondary importance. In many cases ψ is scalar or individual components of a vector ψ may be treated one at a time. Occasionally it is more convenient to consider ψ as a constraint on the full vector θ rather than a specific component; this extension is discussed in Section 8.5.3.

A quite general likelihood-based approach to eliminating the nuisance parameter λ is to replace it by the *constrained maximum likelihood estimate* $\widehat{\lambda}_\psi$ obtained by maximizing $\ell(\psi, \lambda)$ with respect to λ for fixed ψ, and then summarizing knowledge about ψ through the *profile log likelihood*

$$\ell_{\mathrm{p}}(\psi) = \max_\lambda \ell(\psi, \lambda) = \ell(\psi, \widehat{\lambda}_\psi).$$

The observed information function for the profile log likelihood

$$j_{\mathrm{p}}(\psi) = -\frac{\partial^2 \ell_{\mathrm{p}}(\psi)}{\partial \psi \partial \psi^{\mathrm{T}}}$$

can be expressed in terms of the observed likelihood function for the full log likelihood by means of the identity

$$j_p(\psi) = \{j^{\psi\psi}(\psi, \widehat{\lambda}_\psi)\}^{-1},$$

where $j^{\psi\psi}(\psi, \lambda)$ is the (ψ, ψ) block of the inverse of the observed information matrix $j(\psi, \lambda)$. If ψ is a scalar then

$$j_p(\psi) = \frac{|j(\psi, \widehat{\lambda}_\psi)|}{|j_{\lambda\lambda}(\psi, \widehat{\lambda}_\psi)|},$$

where $|\cdot|$ indicates determinant. To a first order of approximation we can treat $\ell_p(\psi)$ as an ordinary log likelihood, for instance basing confidence intervals for ψ on a chi-squared approximation to the likelihood ratio statistic

$$w_p(\psi) = 2\left\{\ell_p(\widehat{\psi}) - \ell_p(\psi)\right\},$$

or on normal approximation to appropriate versions of approximate pivots r, s and t. However, such approximations can give poor results, particularly if the dimension of λ is high. The difficulty is that $\ell_p(\psi)$ is not in general the logarithm of a density function.

It is possible to define a modified likelihood root r^*, analogous to the modified likelihood root of the previous section, that incorporates both an improved approximation and an adjustment for the elimination of the nuisance parameters. This modified likelihood root is central to the higher order approximations used in the following chapters. It will be seen to be well approximated in distribution by the standard normal distribution, and to combine the likelihood root $r(\psi)$, now defined in terms of the profile log likelihood as $\text{sign}(\widehat{\psi} - \psi)w_p^{1/2}(\psi)$, with an approximate pivot $q(\psi)$. In general the form of $q(\psi)$ is somewhat complex, and detailed consideration is postponed until Chapter 8. However, its ingredients are particularly simple when the log likelihood has exponential family form

$$\ell(\psi, \lambda) = \psi u + \lambda^T v - c(\psi, \lambda), \tag{2.8}$$

where the *canonical parameter* is (ψ, λ) and the *natural observation* is (u, v). In such models the distribution of

$$r^*(\psi) = r(\psi) + \frac{1}{r(\psi)}\log\left\{\frac{q(\psi)}{r(\psi)}\right\}$$

is well approximated by the standard normal law, with likelihood root and Wald pivot

$$r(\psi) = \text{sign}(\widehat{\psi} - \psi)\left[2\left\{\ell_p(\widehat{\psi}) - \ell_p(\psi)\right\}\right]^{1/2},$$

$$t(\psi) = j_p^{1/2}(\widehat{\psi})(\widehat{\psi} - \psi)$$

and

$$q(\psi) = t(\psi)\,\rho(\psi, \widehat{\psi}). \tag{2.9}$$

This is a modified form of Wald pivot, where

$$\rho(\psi, \widehat{\psi}) = \left\{ \frac{|j_{\lambda\lambda}(\widehat{\theta})|}{|j_{\lambda\lambda}(\widehat{\theta}_{\psi})|} \right\}^{1/2}$$

and we have introduced the shorthand notation $\widehat{\theta} = (\widehat{\psi}, \widehat{\lambda})$ and $\widehat{\theta}_{\psi} = (\psi, \widehat{\lambda}_{\psi})$.

It can sometimes be useful to decompose the modified likelihood root as

$$r^*(\psi) = r(\psi) + r_{\mathrm{INF}}(\psi) + r_{\mathrm{NP}}(\psi), \tag{2.10}$$

where r_{INF} makes an adjustment allowing for non-normality of r, and r_{NP} compensates r for the presence of the nuisance parameters; expressions for these terms are given in Section 8.6. Graphs of r_{INF} and r_{NP} can be useful in diagnosing the causes of any strong divergences between inferences based on r and on r^*.

Depending on the form of the model $f(y; \psi, \lambda)$, a more direct approach to elimination of nuisance parameters may be available. The most common instances of this are a *conditional likelihood* or a *marginal likelihood*, taken to be the first terms of the decompositions

$$f(y; \psi, \lambda) = \begin{cases} f_{\mathrm{c}}(y \mid u; \psi) f_{\mathrm{m}}(u; \psi, \lambda), \\ f_{\mathrm{m}}(u; \psi) f_{\mathrm{c}}(y \mid u; \psi, \lambda), \end{cases}$$

respectively, when such a factorization is possible. Typically conditional likelihood arises in exponential family models and marginal likelihood arises in transformation models, and numerous examples of both will be given in later chapters. Likelihood roots and related quantities can be defined using such a function, if it is available, and may be used for exact inference on ψ simply by applying the discussion in Section 2.2 to the conditional or marginal likelihood function.

It turns out that the corresponding approximations for conditional and marginal log likelihoods, and the higher order approximations based on r^*, are both closely related to the use of an *adjusted profile log likelihood*

$$\ell_{\mathrm{a}}(\psi) = \ell(\psi, \widehat{\lambda}_{\psi}) - \tfrac{1}{2} \log \left| j_{\lambda\lambda}(\psi, \widehat{\lambda}_{\psi}) \right|,$$

which may be derived from a log posterior marginal density for ψ. Unfortunately ℓ_{a} is not invariant to reparametrization, and a further, in general rather complicated, term must be added to achieve the desired invariant *modified profile log likelihood* $\ell_{\mathrm{m}}(\psi)$. In most of the cases in this book this additional term simplifies considerably; see Sections 8.5.3 and 8.6. An alternative is to use ℓ_{a} but in an *orthogonal parametrization*, chosen so that the contribution made by this extra term is reduced. This implies setting the model up so that the corner of the expected information matrix corresponding to ψ and λ is identically zero. As a first step in investigating the effect of the estimation of nuisance parameters on inference, it can be useful to compare plots of $\ell_{\mathrm{p}}(\psi)$ and $\ell_{\mathrm{a}}(\psi)$.

Illustration: Gamma data

Suppose that a sample $y_1, \ldots, y_n$ is available from the gamma density

$$f(y; \psi, \lambda) = \frac{\lambda^\psi y^{\psi-1}}{\Gamma(\psi)} \exp(-\lambda y), \quad y > 0, \lambda, \psi > 0,$$

and that interest is focused on the shape parameter ψ. The log likelihood may be written as

$$\ell(\psi, \lambda) = \psi u + \lambda v + n\{\psi \log \lambda - \log \Gamma(\psi)\}, \quad \lambda, \psi > 0,$$

where $u = \sum \log y_i$ and $v = -n\bar{y}$, and so $j_{\lambda\lambda}(\psi, \lambda) = n\psi/\lambda^2$. Now $\widehat{\lambda}_\psi = \psi/\bar{y}$, giving $\rho(\psi, \widehat{\psi}) = (\psi/\widehat{\psi})^{1/2}$,

$$\ell_p(\psi) = \psi(u - n) + n\{\psi \log(\psi/\bar{y}) - \log \Gamma(\psi)\}, \quad \psi > 0,$$

and $j_p(\psi) = n\{\Psi'(\psi) - 1/\psi\}$, where $\Psi(\psi)$ denotes the digamma function.

Suppose that the five observations 0.2, 0.45, 0.78, 1.28 and 2.28 are available. The left-hand panel of Figure 2.3 shows both the profile log likelihood for ψ and the conditional log likelihood for the distribution of u given v, which gives exact inferences on ψ because it does not involve λ. The right-hand panel shows the likelihood root $r(\psi)$, the Wald pivot $t(\psi) = j_p(\widehat{\psi})^{1/2}(\widehat{\psi} - \psi)$, and the modified likelihood root $r^*(\psi)$. As with the adjusted profile log likelihood, r^* produces intervals which are shifted towards the origin relative to those based on ℓ_p and on the likelihood root r. The value of $r^*(\psi)$ is extremely close to the curve corresponding to an exact pivot available in this case; the difference is not

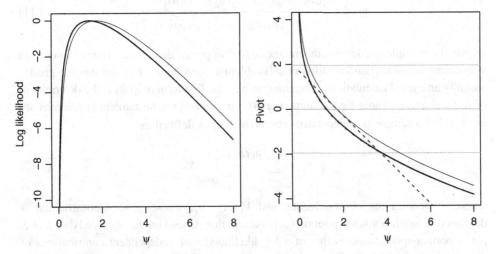

Figure 2.3 Inference for shape parameter ψ of gamma sample of size $n = 5$. Left: profile log likelihood ℓ_p (solid) and the log likelihood from the conditional density of u given v (heavy). Right: likelihood root $r(\psi)$ (solid), Wald pivot $t(\psi)$ (dashes), modified likelihood root $r^*(\psi)$ (heavy), and exact pivot overlying $r^*(\psi)$. The horizontal lines are at $0, \pm 1.96$.

visible in Figure 2.3. Intervals based on the Wald pivot are again unsuitable owing to a poor normal approximation to the distribution of $\widehat{\psi}$. The adjusted profile log likelihood function $\ell_a(\psi)$, computed using the orthogonal parametrizaton $(\psi, \mu = \psi/\lambda)$, is in this model identically equal to the likelihood from the exact conditional density of u given v, although this will rarely be the case, even in exponential family models.

2.4 Further remarks

The most difficult aspect of higher order approximation is determining the approximate pivot $q(\psi)$ to use in the construction of r^*. In linear exponential families, where the parameter of interest is a component of the canonical parameter, the expression (2.9) is readily computable from the fitting of any of the standard generalized linear models, and can be shown in general to lead to excellent approximation of the exact conditional distribution. This is the basis for the `cond` package in the `hoa` bundle for `R`, and is illustrated in Chapters 3 and 4.

There is a similar direct expression for $q(\psi)$ for inference about β_j or σ in regression-scale models, described in more detail in Chapter 8, and illustrated in Chapters 5 and 6; q is a combination of the score pivot s and a correction for nuisance parameters. The r^* pivot in this case gives very accurate approximation to the appropriate marginal distribution. The `marg` package in the `hoa` bundle implements this.

The ideas above have been extended to more general models in numerous ways. One uses a local exponential family approximation in which the canonical parameter is $\varphi(\theta)$, yielding the expression

$$q(\psi) = \frac{|\varphi(\widehat{\theta}) - \varphi(\widehat{\theta}_\psi) \quad \varphi_\lambda(\widehat{\theta}_\psi)|}{|\varphi_\theta(\widehat{\theta})|} \frac{|j(\widehat{\theta})|^{1/2}}{|j_{\lambda\lambda}(\widehat{\theta}_\psi)|^{1/2}}, \tag{2.11}$$

where, for example, φ_θ denotes the matrix $\partial\varphi/\partial\theta^{\mathrm{T}}$ of partial derivatives. Expression (2.11) is invariant to affine transformation of φ (Problem 4, see Chapter 10), and this can greatly simplify analytical calculations. The numerator of the first term of $q(\psi)$ is the determinant of a $d \times d$ matrix whose first column is $\varphi(\widehat{\theta}) - \varphi(\widehat{\theta}_\psi)$ and whose remaining columns are $\varphi_\lambda(\widehat{\theta}_\psi)$. For a sample of independent observations φ is defined as

$$\varphi(\theta)^{\mathrm{T}} = \sum_{i=1}^{n} \left.\frac{\partial\ell(\theta; y)}{\partial y_i}\right|_{y=y^0} V_i, \tag{2.12}$$

where y^0 denotes the observed data, and $V_1, \ldots, V_n$, whose general construction is described in Section 8.4.3 is a set of $1 \times d$ vectors that depend on the observed data alone. An important special case is that of a log likelihood with independent contributions of curved exponential family form,

$$\ell(\theta) = \sum_{i=1}^{n} \{\alpha_i(\theta)y_i - c_i(\theta)\}, \tag{2.13}$$

for which

$$\varphi(\theta)^{\mathrm{T}} = \sum_{i=1}^{n} \alpha_i(\theta) V_i.$$

Affine invariance implies that if $n = d$, that is, we have a reduction by sufficiency to a set of d variables, we can take $\varphi(\theta)^{\mathrm{T}} = (\alpha_1(\theta), \ldots, \alpha_d(\theta))$. Likewise, if $\alpha_i(\theta) \equiv \alpha(\theta)$, then $\varphi(\theta) = \alpha(\theta)$.

The vectors V_i implement conditioning on an approximately ancillary statistic. When y_i has a continuous distribution, V_i is computed by differentiating y_i with respect to θ, for a fixed pivotal z_i; see Section 8.4.3. The expression for V_i computed this way is, as at (8.19),

$$V_i = \frac{\mathrm{d}y_i}{\mathrm{d}\theta^{\mathrm{T}}}\bigg|_{\theta=\widehat{\theta}} = -\left(\frac{\partial z_i}{\partial y_i}\right)^{-1}\left(\frac{\partial z_i}{\partial \theta^{\mathrm{T}}}\right)\bigg|_{\theta=\widehat{\theta}}. \tag{2.14}$$

Inference using (2.11) is relatively easily performed. If functions are available to compute the log likelihood $\ell(\theta)$ and the constructed parameter $\varphi(\theta)$, then the maximisations needed to obtain $\widehat{\theta}$ and $\widehat{\theta}_\psi$, and the derivatives needed to compute (2.11) may be obtained numerically. Explicit formulae for the quantities involved are available for a variety of common classes of models, such as linear regression with non-normal errors and nonlinear regression with normal errors, and these are discussed in later chapters. In many cases standard fitting routines can be exploited and so a negligible additional effort is needed for higher order inference.

Other readily implemented variants of $q(\psi)$ have been proposed. In particular we also use one described in Section 8.5.3 which we call Skovgaard's approximation.

Above, we have supposed that the observations are continuous. If they are discrete, then the ideas remain applicable, but the theoretical accuracy of the approximations is reduced, and the interpretation of significance functions such as $\Phi\{r^*(\theta)\}$ changes slightly; see Sections 3.3, 3.4 and 8.5.4. The construction (2.14) is replaced with

$$V_i = \frac{\mathrm{d}E(y_i;\theta)}{\mathrm{d}\theta^{\mathrm{T}}}\bigg|_{\theta=\widehat{\theta}}; \tag{2.15}$$

see Section 8.5.4 and Problem 55.

There is a close link with analytical approximations useful for Bayesian inference. Suppose that posterior inference is required for ψ and that the chosen prior density is $\pi(\psi, \lambda)$. Then it turns out that using

$$q(\psi) = -\ell_{\mathrm{p}}'(\psi) j_{\mathrm{p}}(\widehat{\psi})^{-1/2} \rho^{-1}(\psi, \widehat{\psi}) \frac{\pi(\widehat{\theta})}{\pi(\widehat{\theta}_\psi)}, \tag{2.16}$$

where ℓ_{p}' is the derivative of $\ell_{\mathrm{p}}(\psi)$ with respect to ψ, in the definition of the modified likelihood root, leads to a Laplace-type approximation to the marginal posteror distribution for ψ; see Section 8.7.

The above sketch of likelihood approximations is the minimum needed to follow the examples in subsequent chapters. Chapter 8 contains a fuller treatment.

Bibliographic notes

The basic definitions and theory of likelihood based inference are described in many textbooks on statistical theory and application; see for example Cox and Hinkley (1974, Chapter 2). The construction of suitable models for particular applications is a difficult and demanding exercise: an illuminating discussion of the issues is given in Chapter 4 of Cox and Snell (1981). Inference based on significance functions or sets of confidence intervals is emphasized in Fraser (1991), Barndorff-Nielsen and Cox (1994) and Efron (1993). The hoa bundle is described in Brazzale (2005); see also Chapter 9. The extensive literature on higher order approximations is surveyed in the bibliographic notes for Chapter 8.

The decomposition of r^* into r_{INF} and r_{NP} was suggested in Pierce and Peters (1992); see also Brazzale (2000, Section 3.2). The cond and marg packages of the hoa bundle provide this decomposition in both the summary and plot methods. For a general expression for r_{INF} and r_{NP} based on q_1 of (8.32), see Barndorff-Nielsen and Cox (1994, Section 6.6.4).

3

Simple illustrations

3.1 Introduction

In this chapter we illustrate the ideas from Chapter 2 on some simple examples, in order to show the calculations in situations where the formulae are available explicitly, so the interested reader can follow the computations and duplicate them easily. In Section 3.2 we consider a single observation from the Cauchy distribution, where first order results are very different from the third order results, and the latter are surprisingly accurate. In Section 3.3 we consider a single Poisson observation; this model has been used in experimental particle physics in efforts to detect a signal in the presence of background. Section 3.4 describes different ways to measure association in a 2×2 contingency table. In Section 3.5 we discuss a two-sample problem, in which the goal is to compare means of rather skewed data under two treatments. Exact results are compared to the higher order approximations, and a brief comparison with bootstrap inference is given.

3.2 Cauchy distribution

As a simple example of a location model, we consider the Cauchy density

$$f(y; \theta) = \frac{1}{\pi\{1 + (y - \theta)^2\}}, \quad -\infty < y < \infty, \tag{3.1}$$

where the location parameter θ can take any real value. The log likelihood function based on a sample $y_1, \ldots, y_n$ of n independent observations is

$$\ell(\theta; y) = -\sum_{i=1}^{n} \log\{1 + (y_i - \theta)^2\},$$

the score equation is

$$\sum_{i=1}^{n} \frac{2(y_i - \widehat{\theta})}{1 + (y_i - \widehat{\theta})^2} = 0,$$

and the observed Fisher information is

$$j(\widehat{\theta}) = 2\sum_{i=1}^{n} \frac{1-(y_i-\widehat{\theta})^2}{\{1+(y_i-\widehat{\theta})^2\}^2}.$$

For $n = 1$ the exact distribution of $\widehat{\theta} = y$ is obtained from (3.1) as $F(y; \theta) = \pi^{-1}\arctan(y - \theta)$, and the likelihood root, score statistic and Wald statistic are, respectively,

$$r(y, \theta) = \text{sign}(y-\theta)[2\log\{1+(y-\theta)^2\}]^{1/2}, \tag{3.2}$$

$$s(y, \theta) = j(\widehat{\theta})^{-1/2}\ell'(\theta) = \sqrt{2}(y-\theta)/\{1+(y-\theta)^2\}, \tag{3.3}$$

$$t(y, \theta) = j(\widehat{\theta})^{-1/2}(\widehat{\theta}-\theta) = \sqrt{2}(y-\theta). \tag{3.4}$$

Here we have emphasized the dependence of the likelihood summaries on both the sample y and the parameter θ; expressions (3.2) – (3.4) are statistics only when θ is fixed at a particular value. In Figure 3.1 we set $\theta = 0$, plot the distribution function $F(y; \theta)$ as a function of y, and compare this to the first order approximations $\Phi\{r(y, 0)\}$, $\Phi\{s(y, 0)\}$ and $\Phi\{t(y, 0)\}$, which are the standard normal approximation to (3.2)–(3.4). We also show the third order approximation $\Phi\{r^*(y, 0)\}$ obtained by taking (2.5) and its variant, the Lugannani–Rice formula (2.6). As this is a location model, the quantity q in these approximations is the score statistic (3.3). The third order approximations are strikingly accurate for so small a sample, despite the facts that the first order approximations are very poor and that the score function is not monotonic in y. Selected numerical values given in Table 3.1 confirm that in this case

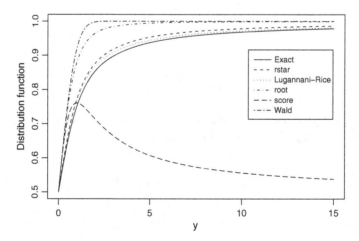

Figure 3.1 The exact distribution function, and the first order and third order approximations, for a sample of size $n = 1$ from the Cauchy distribution, with $\theta = 0$. Only the right-hand side of the distribution function is shown.

Table 3.1 *The exact distribution function for a single observation from the Cauchy distribution with $\theta = 0$, with three first order and two third order approximations.*

y	0.1	1	3	5	10	15	25
$\Phi\{s(y)\}$	0.556	0.760	0.664	0.607	0.555	0.537	0.522
$\Phi\{t(y)\}$	0.556	0.921	0.999	1.000	1.000	1.000	1.000
$\Phi\{r(y)\}$	0.556	0.880	0.984	0.995	0.999	1.000	1.000
$\Phi\{r^*(y)\}$	0.535	0.772	0.918	0.953	0.979	0.987	0.993
Lugannani–Rice	0.535	0.768	0.909	0.944	0.972	0.981	0.989
Exact	0.532	0.750	0.897	0.937	0.968	0.979	0.987

Code 3.1 Code to compare first order and third order approximations for a Cauchy observation.

```
score <- function(y) sqrt(2) * (y-theta)/(1+(y-theta)^2)
root <- function(y) sign(y-theta) * sqrt( 2*(log(1+(y-theta)^2)) )
wald <- function(y) sqrt(2) * (y-theta)
lugr <- function(y) pnorm( root(y) ) +
                dnorm( root(y) ) * ( 1/root(y) - 1/score(y) )
rstar <- function(y) pnorm( root(y) + log(score(y)/root(y))/root(y) )

y <- seq(from = 0, to = 15, by = 0.1)
theta <- 0

plot(y, pnorm(wald(y)), xlab = "y", ylab = "Distribution function",
     type = "l", lty = 6)
lines(y, pnorm(score(y)), lty = 5)
lines(y, pnorm(root(y)), lty = 4)
lines(y, lugr(y), lty = 3)
lines(y, rstar(y), lty = 2)
lines(y, pcauchy(y), lty = 1)
legend(9, 0.9, c("Exact", "rstar", "Lugannani-Rice", "root", "score",
               "Wald"), lty = c(1,2,3,4,5,6) bty = "n")
```

the Lugannani–Rice formula is slightly more accurate than is $\Phi(r^*)$, but even with $n = 1$ both are close enough to the true distribution function for many practical purposes.

The R code to generate Figure 3.1 is given in Code 3.1.

3.3 Top quark

In 1995 the Fermi National Accelerator Laboratory announced the discovery of the top quark, the last of six quarks predicted by the 'standard model' of particle physics. Two simultaneous publications in *Physical Review Letters* described the experiments and the results of the analysis. Evidence for the existence of the top quark was based on signals from the decay of the top quark/antiquark pair created in proton/antiproton collisions. The analysis reported in one of the papers (Abe *et al.*, 1995) was based on a Poisson model:

$$f(y; \theta) = \theta^y e^{-\theta}/y!, \quad y = 0, 1, 2, \ldots, \quad \theta > b.$$

Here b is a positive background count of decay events, and we write $\theta = \mu + b$. Evidence that $\mu > 0$ would favour the existence of the top quark, if the underlying assumptions are correct. Table I of Abe *et al.* (1995) gives the observed values of y as $y^0 = 27$ and of b as 6.7. In fact b is also estimated, but with sufficient precision that this minor complication may be ignored here.

Under the above model the basis of a test of the null hypothesis that $\mu = 0$ against the alternative $\mu > 0$ is whether the observed value y^0 is consistent with a Poisson distribution having mean $\theta = b$. Figure 3.2 shows this distribution function, which is of course a step function, with the smooth approximations $\Phi\{r^*(y, b)\}$ and $\Phi\{r^*(y + \frac{1}{2}, b)\}$ obtained using (2.5). In this exponential family model with canonical parametrization we take q to be the Wald statistic, and $r^*(y, \theta)$ is computed using

$$r(y, \theta) = \text{sign}(y - \theta)[2\{\widehat{\theta} \log(\widehat{\theta}/\theta) - (\widehat{\theta} - \theta)\}]^{1/2},$$

$$t(y, \theta) = \widehat{\theta}^{1/2} \log(\widehat{\theta}/\theta),$$

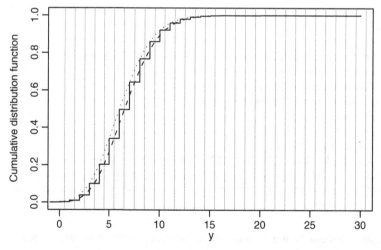

Figure 3.2 Cumulative distribution function for Poisson distribution with parameter 6.7 (solid), with approximations $\Phi\{r^*(y)\}$ (dashes) and $\Phi\{r^*(y + 1/2)\}$ (dots). The vertical lines are at $0.5, 1.5, 2.5, \ldots$

Table 3.2 *Exact upper, lower and mid-P significance levels compared with*
$1 - \Phi(r^*)$, *continuity-corrected versions in which* y^0 *is replaced by* $y^0 \pm \frac{1}{2}$,
and with two first order normal approximations to the distribution of $\widehat{\theta}$.

	$\Pr(Y \geq y^0; \mu = 0)$	$\Pr(Y > y^0; \mu = 0)$
Exact	29.83×10^{-10}	7.06×10^{-10}
Continuity correction	32.36×10^{-10}	7.69×10^{-10}
Mid-P	18.45×10^{-10}	
r^*	15.85×10^{-10}	
$N(\theta, \theta)$	4.45×10^{-4}	2.21×10^{-5}
$N(\theta, \widehat{\theta})$	1.02×10^{-4}	4.68×10^{-5}

where $\widehat{\theta} = y$. The graph shows that for positive integer values of y, the quantities $\Phi\{r^*(y + \frac{1}{2}, b)\}$ and $\Phi\{r^*(y, b)\}$ are close approximations to the cumulative distribution function and the mid-P-value

$$\Pr(Y \leq y; b), \quad \Pr(Y < y; b) + \tfrac{1}{2}\Pr(Y = y; b),$$

respectively. In fact the relative error in these approximations has order n^{-1}, and so they are accurate to second order. Although the sample here consists of a single observation, the structure of the Poisson model means that the parameter θ here plays the role of sample size n. In later examples with discrete data we mostly use the higher order methods without continuity correction, and thereby approximate the mid-P-value, which is preferred by many authors.

To compute the evidence against the null hypothesis in favour of the alternative $\mu > 0$ in this discrete setting we may compute exact upper and lower significance levels, corres-

$$\Pr(Y > y^0; \theta), \quad \Pr(Y \geq y^0; \theta),$$

ponding to evaluated for $y^0 = 27$ and $\theta = b = 6.7$. These values are given in Table 3.2, with approximations obtained by taking the continuity-corrected P-values $1 - \Phi\{r^*(y^0 \pm 1/2; b)\}$, the mid-P-value, and its approximation $1 - \Phi\{r^*(y^0; b)\}$. The relative error of the approximations is below 10% for the first two and around 14% for mid-P – astonishing accuracy for computation of probabilities of order 10^{-9}. Under this model the data give overwhelmingly strong evidence that $\mu > 0$, and thus that there is a signal event different from the background.

Using the normal approximation $y \overset{\cdot}{\sim} N(\theta, \theta)$ gives first order P-values as $1 - \Phi\{(y^0 - \theta)/\sqrt{\theta}\}$, or $1 - \Phi\{(y^0 - \theta)/\sqrt{\widehat{\theta}}\}$ if we use a studentized version. Table 3.2 compares these two normal approximations, which are very different, to the various exact and higher order P-values for testing $\mu = 0$. Both differ from the exact values by several orders of magnitude.

Figure 3.3 shows how the significance functions $\Pr(Y > y^0; b + \mu)$ and $\Pr(Y \geq y^0; b + \mu)$ corresponding to the upper and lower P-values depend on μ, and compares the significance

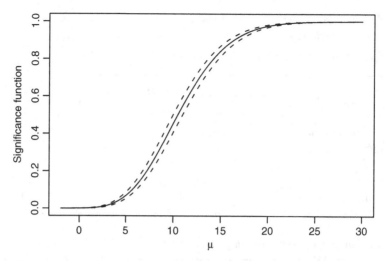

Figure 3.3 Exact upper and lower P-value functions for the top quark parameter μ (dashes), mid-P-value, and the approximation $\Phi(r^*)$ (solid), which coincides with the mid-P-value to within plotting accuracy.

function for mid-P with its approximation $1 - \Phi\{r^*(y^0; b + \mu)\}$. As one would expect from previous discussion, the approximation is essentially exact for any value of μ likely to be of practical interest. The required R code is given in Code 3.2.

Code 3.2 R code to compute exact and third order approximate significance levels for a Poisson observation.

```
root <- function(th, y) { sign(y-th) * sqrt( 2*(y*log(y/th)-y+th) ) }
wald <- function(th, y) { sqrt(y)*log(y/th) }
lugr <- function(th, y = 27) {
  pnorm( root(th, y) ) +
    dnorm( root(th, y) ) * ( 1/root(th, y) - 1/wald(th, y) ) }
rstar <- function(th, y = 27) {
  pnorm( root(th, y) + log( wald(th, y)/root(th, y) )/root(th, y) ) }

b <- 6.7
mu <- seq(from = -2, to = 40, by = 0.01)
y0 <- 27

# Create Figure 3.3

plot(mu, ppois(y0, b+mu, lower.tail = F),
     type = "n", xlab = "mu", ylab = "Significance function")
lines(mu, ppois(y0-1, b+mu, lower.tail = F), lty = 2)
lines(mu, ppois(y0, b+mu, lower.tail = F), lty = 2)
```

```
lines(mu, ppois(y0, b+mu, lower.tail = F) + 0.5 * dpois(y0, b+mu),
      lty = 3)
lines(mu, 1 - rstar(b+mu))

# Compute significance levels

ppois( y0-1, b, lower.tail = F )
1 - rstar(b, y0 - 0.5 )
ppois( y0, b, lower.tail = F )
1 - rstar(b, y0 + 0.5)
ppois( y0, b, lower.tail = F ) + 0.5 * dpois(y0, b)
1 - rstar(b)
pnorm( (y0-1-b)/sqrt(b), lower.tail = F )
pnorm( (y0-b)/sqrt(b), lower.tail = F )
pnorm( (y0-1-b)/sqrt(y0), lower.tail = F )
pnorm( (y0-b)/sqrt(y0), lower.tail = F )

# Create Figure 3.2

y <- seq(from = -1, to = 30, by = 1)
plot(y, ppois(y, b), type = "n", xlim = c(0, 30), xlab = "y",
     ylab = "Cumulative distribution function")
abline(v = y + 1.5, col = "gray")
lines(y, ppois(y, b), type = "s")
x <- seq(from = 0, to = 30, length = 1000)
lines(x, rstar(b, x), lty = 2)
lines(x, rstar(b, x+0.5), lty = 3)
```

3.4 Astronomer data

Cross-classified data arise often in applications, particularly in the social and medical sciences, and one of their commonest forms is the 2×2 table, in which individuals are grouped according to two dichotomous attributes. Table 3.3 gives an example concerning employment of men and women at the Space Telescope Science Institute at Baltimore, which manages the Hubble Space Telescope. Five of the seven women scientists hired in the period 1998–2002 left within that period, while only one of 19 men hired in that period did so. The mismatch in the proportions 5/7 and 1/19 raised questions about the workplace environment for women.

Very often with data in a 2×2 table one of the sets of categories is regarded as fixed beforehand, while the other is treated as a random outcome. Then the simplest model is that the random outcome is a sum of independent identically distributed Bernoulli variables and so is binomial. Here this must be treated with some skepticism, as it seems unlikely that individual resignations were independent. Leaving this aside, in the present context the number of men leaving could be modelled as a binomial variable Y_0, with probability p_0 and m_0 the number of men hired, while the corresponding quantities for the women are Y_1, p_1, and m_1.

The usual purpose of analysis is to compare p_0 and p_1, and this can be performed on various scales. Some possibilities are the difference of log odds, the probability ratio and

Table 3.3 *Employment of men and women at the Space Telescope Science Institute, 1998–2002 (Science, **299**, 993, 14 February 2003).*

	Left	Stayed	Total
Men	1	18	19
Women	5	2	7
Total	6	20	26

the difference of probabilities,

$$\psi = \log\left(\frac{p_1}{1-p_1}\right) - \log\left(\frac{p_0}{1-p_0}\right), \quad \xi = p_1/p_0, \quad \chi = p_1 - p_0.$$

The first of these is a canonical parameter in the exponential family model

$$f(y_0; p_0)f(y_1; p_1) = \binom{m_1}{y_1}p_1^{y_1}(1-p_1)^{m_1-y_1}\binom{m_0}{y_0}p_0^{y_0}(1-p_0)^{m_0-y_0}, \tag{3.5}$$

as we see on rewriting this probability density as

$$\binom{m_1}{y_1}\binom{m_0}{y_0}\exp\left\{y_1\psi + (y_1 + y_0)\lambda - m_0\log(1+e^\lambda) - m_1\log(1+e^{\lambda+\psi})\right\}$$

and comparing it to the log likelihood (2.8). The parameters λ and ψ take real values and are variation independent, and

$$p_0 = \frac{e^\lambda}{1+e^\lambda}, \quad p_1 = \frac{e^{\lambda+\psi}}{1+e^{\lambda+\psi}}.$$

Comparison using the difference of log odds is natural from a theoretical point of view, and moreover the interpretation of ψ is the same whether the data result from a prospective or a retrospective study; this can be important in epidemiological work. The difference and the ratio of probabilities have direct interpretations and are more readily grasped by non-experts, but are more awkward to deal with from a mathematical point of view.

Suppose initially that the log odds ratio ψ is the parameter of interest. As ψ and λ are canonical parameters in the linear exponential family, conditioning on the statistic $V = Y_0 + Y_1$ associated with λ yields exact inference for ψ based on the extended hypergeometric density of Y_1 given V. A conditional likelihood for ψ may be obtained from this density,

$$f(y_1 \mid v; \psi) = \frac{\binom{m_1}{y_1}\binom{m_0}{v-y_1}e^{\psi y_1}}{\sum_{u=y_-}^{y_+}\binom{m_1}{u}\binom{m_0}{v-u}e^{\psi u}}, \quad y_1 = y_-, \ldots, y_+; \tag{3.6}$$

here $y_- = \max(0, v - m_0)$, $y_+ = \min(m_1, v)$ and $v = y_0 + y_1$ is the observed value of V. A significance level for testing $\psi = 0$, i.e. that men and women have equal resignation probabilities, against the alternative $\psi > 0$, i.e. that women are more likely to leave than men, is

$$p_{\text{obs},-} = \Pr(Y_1 \leq y_1 \mid V = v; \psi)|_{\psi=0} = \sum_{u=y_-}^{y_1} f(u \mid v; 0); \qquad (3.7)$$

the mid-P-value is

$$p_{\text{mid},-} = \Pr(Y_1 < y_1 \mid V = v; 0) + \tfrac{1}{2} f(y_1 \mid v; 0).$$

The data in Table 3.3 give $p_{\text{obs},-} = 0.001\,76$ and $p_{\text{mid},-} = 0.000\,90$: very strong evidence that women are more likely to leave.

As ψ and λ are the canonical parameters of an exponential family model, approximate inference may be based on the discussion following (2.8). The corresponding computations are automated in the cond package for R; see Code 3.3. Setting $\psi = 0$ gives $r = 3.447$, $t = 2.873$, $q = 2.061$, $r^* = 3.298$ and significance level $1 - \Phi(r^*) = 0.000\,49$. This is not far from the mid-P probability obtained above, while computing r, q and r^* for the adjusted data in which $(y_1, y_0) = (5, 1)$ are replaced by $(y_1 - \tfrac{1}{2}, y_0 + \tfrac{1}{2}) = (4.5, 1.5)$ gives a significance level for r^* of $0.001\,85$, close to the exact value from (3.7).

The left-hand panel of Figure 3.4 compares the profile and adjusted log likelihoods ℓ_p and ℓ_a for the log odds with the log conditional likelihood obtained as the logarithm of (3.6). The adjusted curve agrees closely with the conditional curve, while the profile log likelihood shows how the overall maximum likelihood estimate tends to be biased away from zero in such cases. The right-hand panel shows how the roots r and r^* depend

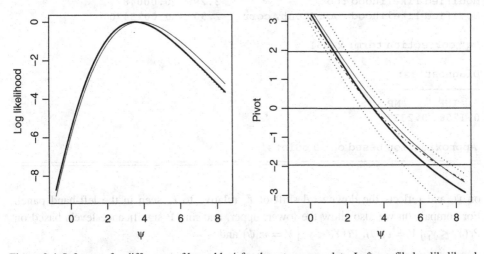

Figure 3.4 Inference for difference of log odds ψ for the astronomer data. Left: profile log likelihood ℓ_p (solid) and its adjusted form ℓ_a (heavy), with exact conditional log likelihood (dots). Right: likelihood root (solid), modified likelihood root (heavy), with mid-P (dot-dash) and upper and lower significance levels (dots). The horizontal lines show the 0.025, 0.5 and 0.975 quantiles of the standard normal distribution.

Code 3.3 Approximate conditional inference for the Hubble data.

```
> hubble <- data.frame( r = c(1, 5), m = c(19, 7), x = c(0, 1) )
> hubble
  r  m x
1 1 19 0
2 5  7 1
> summary( hubble.fit <- glm( cbind(r, m-r) ~ x, binomial,
+                             data = hubble ) )

Coefficients:
             Estimate  Std. Error  z value  Pr(>|z|)
(Intercept)   -2.890       1.027   -2.813    0.00490
x              3.807       1.325    2.873    0.00407

> summary( cond.glm( hubble.fit, offset = x ), test = 0 )

          Estimate  Std. Error
uncond.      3.807      1.325
cond.        3.549      1.241

Test statistics
---------------
 hypothesis : coef( x ) = 0
                                       statistic  tail prob.
Wald pivot                                 2.873  0.0020330
Wald pivot (cond. MLE)                     2.860  0.0021200
Likelihood root                            3.447  0.0002838
Modified likelihood root                   3.298  0.0004877
Modified likelihood root (cont. corr.)     2.903  0.0018470

"q" correction term: 2.061

Diagnostics:
-----------
   INF      NP
0.1256  0.2337

Approximation based on 20 points
```

on ψ, and reflects the downward shift of ℓ_a relative to ℓ_p seen in the left-hand panel. For comparison we also show the lower, upper, and mid-P significance levels based on $\Pr(Y_1 \leq y_1 \mid V = v; \psi)$, $\Pr(Y_1 < y_1 \mid V = v; \psi)$ and

$$\Pr(Y_1 < y_1 \mid V = v; \psi) + \tfrac{1}{2}\Pr(Y_1 = y_1 \mid V = v; \psi).$$

Now suppose that the parameter of interest is the ratio of probabilities $\xi = p_1/p_0$, and that p_0 is taken as the nuisance parameter. Unlike in the discussion above, in this case there is no exact significance level for testing a hypothesis on ξ, so there is no target value with which higher order procedures may be compared. The constraints $0 \leq p_0, \xi p_0 \leq 1$ imply that

$$0 < \xi, \quad 0 \leq p_0 \leq \min(1, \xi^{-1}).$$

For numerical purposes it is best to rewrite the model in terms of $\psi = \log \xi$ and λ, where

$$p_0(\psi, \lambda) = \min(1, e^{-\psi}) \frac{e^\lambda}{1 + e^\lambda}, \quad p_1(\psi, \lambda) = e^\psi p_0(\psi, \lambda), \quad \psi, \lambda \in \mathbb{R}.$$

The log likelihood may be found by substituting these into (3.5). The maximum likelihood estimates are $\widehat{\psi} = 2.608$ and $\widehat{\lambda} = 0.916$, with standard errors 1.002 and 0.837.

The hypothesis of equal resignation probabilities for men and women corresponds to $\psi = 0$. The Wald statistic for $\psi = 0$ is $t = (2.602 - 0)/1.002 = 2.602$, and the value of the likelihood root is $r = 3.447$, giving respective P-values of $1 - \Phi(2.602) = 0.004\,63$ and $1 - \Phi(3.447) = 0.000\,28$. Both give strong evidence that the ratio of probabilities ξ exceeds unity.

Code 3.4 Higher order inference for the ratio for the Hubble data.

```
nlogL <- function(psi, lam, d)
{
  m <- min(1, exp(-psi))
  p0 <- m / (1 + exp(-lam))
  p1 <- p0 * exp(psi)
  -sum( dbinom(d$r, size = d$m, prob = c(p0, p1), log = T) )
}

make.V <- function(th, d) NULL # V is not needed in this case

phi <- function(th, V, d)
{
  psi <- th[1]
  lam <- th[-1]
  m <- min(1, exp(-psi))
  p0 <- m / (1 + exp(-lam))
  p1 <- p0 * exp(psi)
  phi1 <- log( p0/(1-p0) )
  phi2 <- log( p1/(1-p1) )
  c(phi1, phi2)
}

( hubble.fr <- fraser.reid( 0, nlogL, phi, make.V,
                 th.init = c(0, 0.1), hubble ) )
```

In order to compute a higher order significance probability, we could take the approach outlined after (2.11). The model is an exponential family with natural observation and canonical parameter

$$(y_0, y_1), \quad (\log\{p_0/(1-p_0)\}, \log\{p_1/(1-p_1)\}),$$

so (2.15) implies that the required reparametrisation $\varphi(\theta)$ of $\theta = (\psi, \lambda)^{\mathsf{T}}$ may be taken to be

$$\varphi_1(\psi, \lambda) = \log\left\{\frac{p_0(\psi, \lambda)}{1 - p_0(\psi, \lambda)}\right\}, \quad \varphi_2(\psi, \lambda) = \log\left\{\frac{p_1(\psi, \lambda)}{1 - p_1(\psi, \lambda)}\right\}.$$

The code needed for the higher order computations, given in Code 3.4, yields $q = 2.061$, and hence $r^* = 3.298$, giving the P-value $1 - \Phi(r^*) = 0.00049$, as before.

The corresponding computations for the difference of probabilities are left to Problem 9; see also Section 4.4.

3.5 Cost data

Table 3.4 gives the costs of treatment for two groups of patients with a history of deliberate self harm. One group received the usual therapy, and the other received cognitive behavioural therapy. As is typical of medical cost data, the observations are highly skewed. Probability plots suggest that either the exponential distribution or the log normal distribution might provide suitable models. The data were collected to compare the costs of the two treatments, which in statistical terms suggests focusing on either the difference or the ratio of the two group means. Let $X_1, \ldots, X_n$ denote the costs in thousands of pounds for the first group, and $Y_1, \ldots, Y_m$ those for the second group.

Suppose first that the data are modelled as independent exponential variables, with $E(X_i) = \mu_1$ and $E(Y_i) = \mu_2$. If the quantity of interest is the ratio of means, $\psi = \mu_1/\mu_2$, then exact confidence intervals may be constructed using the fact that $\sum X_j/\mu_1$ and $\sum Y_i/\mu_2$ have independent chi-squared distributions with $2n$ and $2m$ degrees of freedom. This implies that

$$\left(\frac{\overline{X}}{\mu_1}\right) \Big/ \left(\frac{\overline{Y}}{\mu_2}\right) \quad \sim \quad F_{2n, 2m},$$

where $\overline{X}$ and $\overline{Y}$ are the group averages. The resulting equi-tailed $(1 - 2\alpha)$ confidence interval is

$$\left(\overline{X}/\{\overline{Y}F_{2n,2m}(1-\alpha)\}, \overline{X}/\{\overline{Y}F_{2n,2m}(\alpha)\}\right), \tag{3.8}$$

where $F_{\nu_1, \nu_2}(\alpha)$ denotes the α quantile of the F distribution with ν_1 and ν_2 degrees of freedom. The data in Table 3.4 have $n = 13$, $m = 18$, $\overline{x} = 1.283$ and $\overline{y} = 0.648$, the units for these being thousands of pounds, so $\widehat{\psi} = \overline{x}/\overline{y} = 1.98$, and the 95% equi-tailed confidence interval for ψ is $(0.98, 4.19)$. The new therapy is estimated to be twice as expensive, but the data are just consistent with a true cost ratio of 1.

Table 3.4 *Costs (pounds sterling) of cognitive behavioural therapy (group 1) and the usual therapy (group 2) for patients with a history of deliberate self harm (Evans* et al., *1999).*

Group 1	30	172	210	212	335	489	651	1263	1294
	1875	2213	2998	4935					
Group 2	121	172	201	214	228	261	278	279	351
	561	622	694	848	853	1086	1110	1243	2543

We now illustrate the use of higher order approximations involving (2.11). The joint density of the data is the exponential family

$$\mu_1^{-n}\mu_2^{-m}\exp\left(-\sum_{i=1}^{n}x_i/\mu_1 - \sum_{i=1}^{m}y_i/\mu_2\right), \quad \mu_1, \mu_2 > 0.$$

The ratio of exponential means is not a canonical parameter, so exact conditional inference is not available by conditioning on a component of the sufficient statistic. With $\mu_2 = \lambda$ and $\mu_1 = \psi\lambda$, the canonical parameter is

$$\varphi = (-1/(\psi\lambda), -1/\lambda)^{\mathrm{T}}, \tag{3.9}$$

and, writing $\theta^{\mathrm{T}} = (\psi, \lambda)$, we have

$$\varphi_\theta(\theta) = \frac{\partial\varphi(\theta)}{\partial\theta^{\mathrm{T}}} = \begin{pmatrix} 1/(\psi^2\lambda) & 1/(\psi\lambda^2) \\ 0 & 1/\lambda^2 \end{pmatrix},$$

the second column of which contains φ_λ.

The Fisher information components in terms of θ are

$$j_{\theta\theta}(\widehat\theta) = \begin{pmatrix} n/\widehat\psi^2 & n/(\widehat\lambda\widehat\psi) \\ n/(\widehat\lambda\widehat\psi) & (n+m)/\widehat\lambda^2 \end{pmatrix},$$

$$j_{\lambda\lambda}(\widehat\theta_\psi) = (n+m)/\widehat\lambda_\psi^2,$$

and combining them with (3.9) we obtain

$$q(\psi) = \frac{\left| \varphi(\widehat\theta) - \varphi(\widehat\theta_\psi) \quad \varphi_\lambda(\widehat\theta_\psi) \right|}{\left| \varphi_\theta(\widehat\theta) \right|} \left\{ \frac{\left| j(\widehat\theta) \right|}{\left| j_{\lambda\lambda}(\widehat\theta_\psi) \right|} \right\}^{1/2}$$

$$= \frac{\widehat\psi - \psi}{\widehat\psi\sqrt{(1/n + 1/m)}} \frac{\widehat\psi\widehat\lambda}{\psi\widehat\lambda_\psi}. \tag{3.10}$$

Table 3.5 *Probability left and right of the exact confidence limits for the ratio of exponential means for the cost data. The exact limits were computed using the F distribution and the approximate probability using the r^* approximation. In this example the Lugannani–Rice and r^* approximations are identical.*

Exact	0.20	0.10	0.05	0.025	0.001	0.0001
Approx. (left)	0.199 98	0.099 992	0.049 997	0.024 999	0.001 000	0.000 100
Approx. (right)	0.200 03	0.100 239	0.050 015	0.025 009	0.001 000	0.000 100

The denominator of the first factor in (3.10) is the standard error of $\widehat{\psi}$ computed by the delta method, so the first factor itself is $t(\psi)$, and in this exponential family model $q(\psi)$ is indeed a perturbed Wald statistic; cf. (2.9).

Here the Lugannani–Rice formula and r^* give essentially exact results: the 95% confidence interval for ψ based on r^* is (0.98, 4.185). The accuracy of the r^* approximation for other confidence bounds is illustrated in Table 3.5.

If instead the quantity of interest is the difference $\delta = \mu_1 - \mu_2$, we again set $\lambda = \mu_2$, in terms of which the log likelihood may be written

$$\ell(\delta, \lambda) = -m\{\log(\lambda + \delta) + \bar{y}/(\lambda + \delta)\} - n(\log \lambda + \bar{x}/\lambda), \qquad (3.11)$$

where δ takes values in the real line and $\lambda > \max(0, -\delta)$. The maximum likelihood estimate is $\widehat{\delta} = \bar{x} - \bar{y} = 0.635$ thousand pounds with standard error 0.387, so normal approximation to the distribution of the Wald pivot gives 95% confidence interval $635 \pm 1.96 \times 387 = (-124, 1394)$ pounds for the mean cost difference δ.

In this parametrization the canonical parameter of the exponential family is $\varphi = (-1/(\delta + \lambda), -1/\lambda)$, and computations analogous to those yielding (3.10) are straightforward. With δ fixed, expression (3.11) is readily maximized numerically as a function of λ, and r, t and r^* are shown as functions of the interest parameter δ in the left-hand panel of Figure 3.5. There is a substantial difference between the function corresponding to the Wald pivot (2.3), here $(0.635 - \delta)/0.387$, and those for r and r^*, which are asymmetric about $\widehat{\delta}$. Confidence intervals based on r and r^* are as usual obtained by determining the intersections of its curve and the horizontal lines representing the required quantiles of the standard normal distribution. The 95% intervals for δ based on r and on r^*, $(-23, 1710)$ and $(-22, 1781)$ pounds, reflect the asymmetry of the likelihood root.

Another possible model is that the costs follow log-normal distributions, so that $\log X_1, \ldots, \log X_n$ is a normal random sample with mean μ_1 and variance σ_1^2, and $\log Y_1, \ldots, \log Y_m$ is a normal random sample with mean μ_2 and variance σ_2^2. Then the log likelihood $\ell(\mu_1, \mu_2, \sigma_1^2, \sigma_2^2)$ equals

$$-\frac{1}{2}\left\{n\log \sigma_1^2 + \sum_{i=1}^{n} \frac{(\log x_i - \mu_1)^2}{\sigma_1^2} + m\log \sigma_2^2 + \sum_{i=1}^{m} \frac{(\log y_i - \mu_2)^2}{\sigma_2^2}\right\}, \qquad (3.12)$$

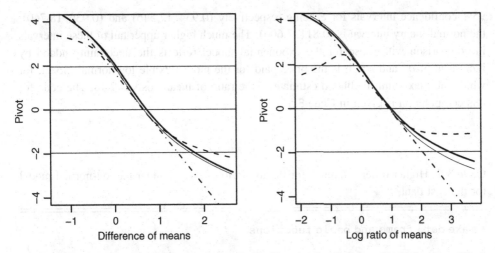

Figure 3.5 Likelihood analysis for cost data. Left: likelihood root r (solid), q (dashes) and modified likelihood root r^* (heavy) as functions of difference of exponential means. The diagonal line (dot-dash) shows the Wald pivot $t(\delta) = (\widehat{\delta} - \delta)/v^{1/2}$, and the horizontal lines show the 0.025, 0.5, and 0.975 quantiles of the standard normal distribution. Confidence intervals are found from their intersections with the curves for the likelihood roots. Right: corresponding quantities for log ratio of log-normal means.

corresponding to an exponential family with canonical parameter

$$\varphi = (\mu_1/\sigma_1^2, 1/\sigma_1^2, \mu_2/\sigma_2^2, 1/\sigma_2^2)^{\mathsf{T}}.$$

The group means are $E(X_i) = \exp(\mu_1 + \sigma_1^2/2)$ and $E(Y_i) = \exp(\mu_2 + \sigma_2^2/2)$, and interest may again focus either on their ratio or on their difference. In each case there is a single parameter of interest, the log ratio

$$\xi = \mu_1 + \sigma_1^2/2 - \mu_2 - \sigma_2^2/2$$

or the difference of log-normal means

$$\exp(\mu_1 + \sigma_1^2/2) - \exp(\mu_2 + \sigma_2^2/2),$$

both of which take values in the real line. Inference for ξ can again be performed using (2.11) , expressing φ in terms of interest and nuisance parameters through

$$\mu_1 = \xi + \lambda_1 + \tfrac{1}{2}e^{2\lambda_3} - \tfrac{1}{2}e^{2\lambda_2}, \quad \mu_2 = \lambda_1, \quad \sigma_1 = e^{\lambda_2}, \quad \sigma_2 = e^{\lambda_3},$$

where the parametrization is in terms of the log variances for numerical stability; parametrization-invariance means that this does not affect inferences for the interest parameter ξ. In this case it is best to obtain the matrices of derivatives in (2.11) numerically. The result is quantities r, q and r^*, whose dependence on ξ is shown in the right-hand panel of Figure 3.5. In this case q is not monotonic, and r and r^* differ considerably,

95% confidence intervals for ξ being respectively $(0.934, 12.546)$ and $(0.936, 18.760)$; the normal-theory interval is $(0.811, 7.606)$. The much higher upper tail of these intervals in comparison with those for the exponential model reflects the uncertainty added by ignorance of σ_1^2 and σ_2^2; it is the price paid for the more flexible log-normal model, for which the maximum likelihood estimate of the ratio of mean costs is 2.48. The code for this computation is given in Code 3.5.

Code 3.5 Higher order inference for the log ratio of means under the log-normal model for the cost data.

```
# make data frame and basic functions

cost <- data.frame(
            f = factor( c(rep(1, 13), rep(2, 18)) ),
            y = c( 30,172,210,212,335,489,651,1263,1294,1875,2213,
                   2998,4935,121,172,201,214,228,261,278,279,351,561,
                   622,694,848,853,1086,1110,1243,2543 ) )

nlogL <- function(psi, lam, d) {
# d = data
 s1 <- exp(lam[2])
 m2 <- lam[1]
 s2 <- exp(lam[3])
 m1 <- psi + m2 + s2^2/2 - s1^2/2
 - sum( dnorm( log(d$y), mean = ifelse(d$f==1, m1, m2),
            sd = ifelse(d$f==1, s1, s2), log = T) )
}

phi <- function(th, V, d) {
 psi <- th[1]
 lam <- th[-1]
 s1 <- exp(lam[2])
 m2 <- lam[1]
 s2 <- exp(lam[3])
 m1 <- psi + m2 + s2^2/2 - s1^2/2
 c( m1/s1^2, 1/s1^2, m2/s2^2, 1/s2^2 )
}

make.V <- function(th, d) NULL

cost.lnorm.rat <- fraser.reid( psi = NULL, nlogL = nlogL, phi = phi,
                        make.V = make.V,
                        th.init = c(0, 5, 2, 5), d = cost )
plot.fr( cost.lnorm.rat )
lik.ci( cost.lnorm.rat )
```

Table 3.6 *Comparison of bootstrap and higher order 95% confidence intervals for ratio and difference of mean costs of treatment (pounds), for cost data. Bootstrap intervals are based on 10 000 simulations.*

Model	Approach	Confidence interval	Ratio	Difference
Nonparametric	Bootstrap	Normal	(0.97, 4.23)	(−160, 1420)
		Student	(0.87, 4.58)	(−90, 1870)
		BC_a	(0.91, 3.93)	(−10, 1610)
Exponential	Bootstrap	Normal	(0.97, 4.16)	(−130, 1400)
		Student	(0.88, 4.70)	(−80, 1840)
	Higher order		(0.98, 4.19)	(−20, 1780)
Log normal	Bootstrap	Normal	(0.59, 5.34)	(−1760, 2390)
		Student	(0.77, 5.66)	(−40, 2280)
	Higher order		(0.81, 18.76)	

Inference for the difference of means involves the implicit formulation described at (8.35), but in this case maximization of (3.12) under the constraint that $\exp(\mu_2 + \sigma_2^2/2) - \exp(\mu_1 + \sigma_1^2/2)$ takes a fixed value seems to be numerically quite unstable, and we were unable to obtain the higher order confidence interval for the difference.

An alternative to higher order asymptotic analysis is the use of resampling procedures such as the bootstrap. There are some important differences, however. Bootstrap calculations involve simulation from a fitted model, either a non-parametric one, in which case the simulation typically involves resampling of the original observations, or a parametric one, such as the exponential or log-normal densities fitted by maximum likelihood estimation. The simulation is rarely conditioned on any ancillary statistic, and Monte Carlo variation may make it computationally expensive to obtain confidence intervals with very high coverage. In this case we performed three bootstrap simulations, using non-parametric simulation from the two groups separately, and parametric simulation from the exponential model and from the log-normal model fitted by maximum likelihood estimation. The estimate for the difference of means was taken to be $\bar{x} - \bar{y}$, with estimated variance $s_1^2/n + s_2^2/m$, where s_1^2 and s_2^2 are the sample variances for the two groups. The log ratio was estimated by $\log(\bar{x}/\bar{y})$, with estimated variance $s_1^2/(n\bar{x}^2) + s_2^2/(n\bar{y}^2)$. The confidence interval methods reported use normal approximation to the distribution of the estimators, with estimated moments, and studentization using bootstrap approximation to the distribution of the t pivot. Illustrative code for such computations is given in Code 3.6.

Table 3.6 compares some resampling-based 95% confidence intervals with those from the higher order procedures described above. There is broad agreement between the more accurate bootstrap procedures, namely the studentized and BC_a methods, and the higher

Code 3.6 Bootstrap inference for the difference and ratio for the cost data.

```
# Basic bootstrapping functions

stat <- function(data) {
# compute difference and ratio of means, and estimated variances
  m <- tapply( data$y, data$f, mean )
  v <- tapply( data$y, data$f, var )
  n <- tapply( data$y, data$f, length )
  c( m[1] - m[2], v[1]/n[1] + v[2]/n[2], log( m[1]/m[2] ),
     v[1]/(m[1]^2 * n[1]) + v[2]/(m[2]^2 * n[2]) )
}

cost.f <- function(data, i = 1:nrow(data)) stat(cost[i,])

library(boot)

# nonparametric bootstrap

boot.ci( cost.boot <- boot( cost, cost.f, R = 10000,
                            strata = cost$f ) ) # for difference
plot(cost.boot)
boot.ci( cost.boot, index = c(3,4) ) # results for log ratio of means
plot(cost.boot, index = 3)

# parametric bootstrap using exponential model, giving results for
# difference of means

cost.mle <- c( tapply(cost$y, cost$f, mean),
               tapply(cost$y, cost$f, length) )
cost.sim <- function(data, mle) {
 d <- data
 d$y <- c( rexp(mle[3], 1/mle[1]), rexp(mle[4], 1/mle[2]) )
 d
}
boot.ci(
  cost.para <- boot( cost, stat, R = 10000, ran.gen = cost.sim,
                   mle = cost.mle, sim = "parametric" ),
    type = c("norm,"stud") )

# parametric bootstrap using log-normal model, giving results for
# difference of means

v <- tapply( log(cost$y), cost$f, var )
n <- tapply( cost$y, cost$f, length )
sig <- (n-1)* v/n
cost.mle.lnorm <- c( tapply(log(cost$y), cost$f, mean), sqrt(sig), n )
cost.sim.lnorm <- function(data, mle ) {
```

```
   d <- data
   d$y <- exp( c( rnorm(mle[5], mean = mle[1], sd = mle[3]),
                  rnorm(mle[6], mean = mle[2], sd = mle[4]) ) )
   d
}
boot.ci(
 cost.para.lnorm <- boot( cost, stat, R = 10000,
                          ran.gen = cost.sim.lnorm,
                          mle = cost.mle.lnorm, sim = "parametric" ),
        type = c("norm, "stud") )
```

order methods. The most striking exception to this is the remarkably high right endpoint of the interval for the ratio under the log-normal model. Inference for this model can be difficult, and the discrepancy presumably arises because although the bootstrap simulations are generated under the fitted model, a wide range of other parameter values, including some very different from the fitted model, would also be compatible with the data.

All the analyses point to the same conclusion: although the new therapy seems more expensive than the usual therapy, the data are consistent with a value of 1 for the ratio of the mean costs and with a value of 0 for the difference, at confidence level 95%.

Bibliographic notes

There are a number of illustrations in the literature of the surprising accuracy of the r^* approximation in simple one parameter models; see for example Reid (1996), Severini (2000a, Chapter 7) and Butler (2007, Chapters 1, 2). The Cauchy model is discussed in Barndorff-Nielsen (1990) and Fraser (1990).

Inference for a bounded Poisson parameter has received considerable attention in the physics literature. Feldman and Cousins (1998) proposed a method for constructing confidence intervals, called the 'unified method', that adjusts between one-sided and two-sided intervals depending on the strength of the signal. This method is described in Mandelkern (2002) and criticised by several statisticians in the discussion. Fraser *et al.* (2004) discuss the use of the P-value function in this model, and the higher order approximation discussed here. They also describe higher order approximation when the background is estimated from auxiliary Poisson data.

Conditioning in 2×2 tables has a long history of animated discussion in the statistics literature. A fairly recent reference is Yates (1984); see also Little (2006). The conditional distribution used here has been criticized on the grounds that the marginal total, which is being conditioned on, is not ancillary for the log odds ratio, and so the distribution of the marginal total might contain information about the parameter of interest as well as about the nuisance parameters. The discussions are complicated by the fact that the conditional distribution has fewer points of support, and hence fewer achievable levels of significance in applications. Our view is that using the continuous approximation, either by mid-P-values or $\Phi(r^*)$, is preferable, and that the benefit of conditioning, exact elimination of nuisance parameters, far outweights the slight loss of information.

The difficulties caused by discreteness in analyzing binomial data are discussed from a different point of view by Brown *et al.* (2001), who show that the coverage properties of confidence intervals based on the Wald pivot are very unsatisfactory, and recommend alternatives. See also Agresti and Coull (1998) and Problems 8 and 24.

General references on bootstrap methods are Davison and Hinkley (1997) and Efron and Tibshirani (1993). The connection between bootstrap procedures, particularly non-parametric bootstrap methods, and parametric approximations of tail areas is not completely clear at the time of writing; an accessible account of what is known is DiCiccio and Efron (1996); see also Lee and Young (2005).

4

Discrete data

4.1 Introduction

In this chapter we consider data with discrete response variables. The most common distributions for such responses are the binomial and the Poisson, used for binary outcomes and for count data, most often in the guise of logistic regression and log-linear models, respectively. These models are exponential families, which have desirable statistical properties and have been widely studied. The approximations outlined in the discussion following (2.8) have been implemented in the cond package for canonical parameters of these models, so that values for r, t and r^*, and the related tail area approximations, are all readily available. The approach mentioned in Section 2.4 may be used for other discrete response models, as illustrated in Sections 3.4, 4.4, and 4.6.

In linear exponential families exact significance levels and confidence intervals are available, at least in principle, through the use of conditioning arguments that allow the elimination of nuisance parameters. To illustrate this, suppose that independent binary responses $Y_1, \ldots, Y_n$ depend on values of a scalar covariate $x_1, \ldots, x_n$ through the logistic regression model

$$\Pr(Y_i = 1) = 1 - \Pr(Y_i = 0) = \frac{\exp(\alpha + \beta x_i)}{1 + \exp(\alpha + \beta x_i)}, \quad i = 1, \ldots, n.$$

Then the joint density of the responses may be expressed as

$$\prod_{i=1}^{n} \Pr(Y_i = 1)^{y_i} \Pr(Y_i = 0)^{1-y_i} = \frac{\exp(\alpha u + \beta v)}{\prod_{i=1}^{n} \{1 + \exp(\alpha + \beta x_i)\}}, \tag{4.1}$$

where $(u, v) = (\sum y_i, \sum x_i y_i)$ is a sufficient statistic; the corresponding log likelihood is of exponential family form (2.8). The conditional density of V given that $U = u$ is therefore

$$f(v \mid u; \alpha, \beta) = \frac{\exp(\alpha u + \beta v)}{\sum' \exp(\alpha u + \beta v')} = \frac{\exp(\beta v)}{\sum' \exp(\beta v')} = f(v \mid u; \beta), \tag{4.2}$$

where $\sum'$ denotes summation over the subset $\mathcal{Y}_u$ of the sample space

$$\mathcal{Y} = \{(y_1, \ldots, y_n) : y_i = 0, 1, i = 1, \ldots, n\}$$

for which $\sum y_i = u$. The denominator of (4.1) cancels from expression (4.2), which does not depend on α. A parallel argument shows that the conditional density of U given V does not depend on β. This is a special case of a general exponential family result, which also applies if one or both of α and β are vectors. It exemplifies the first type of decomposition expressed in Section 2.3, and discussed in more detail in Section 8.6.1. The implication is that inferences on β are available from $f(v \mid u; \beta)$ without reference to the unknown α; a test of the hypothesis $\beta = 0$, for instance, may be based on tail probabilities such as

$$\Pr(V \leq v_{\text{obs}} \mid U = u; \beta)|_{\beta=0} = \sum_{v \leq v_{\text{obs}}} f(v \mid u; \beta)\,|_{\beta=0},$$

where v_{obs} is the observed value of V. The resulting test is exact: no approximation is required, and the significance level does not depend upon α. Likewise confidence intervals for β may be obtained by treating the conditional density (4.2) as a single-parameter likelihood function.

In applications the main difficulty with exact inference based on discrete distributions is that enumeration of the required subset of $\mathcal{Y}$ will often require specialized software. Moreover, the discreteness limits the significance levels that are attainable for a given β; indeed, sometimes the discreteness is so pronounced that useful exact inferences are unobtainable – but despite this higher order procedures may provide sensible results. Furthermore, standard inference procedures may not be useful when the likelihood is maximized on the boundary of the parameter space. The most common sign of this is that certain maximum likelihood estimates and standard errors are effectively infinite, but the fitting algorithm may also signal a failure to converge.

To underline these points, consider an artifical data set with $n = 6$ binary responses, whose values of $x_1, \ldots, x_6$ are $0, 1, 1, 2, 4, 8$. The left-hand panel of Figure 4.1 shows

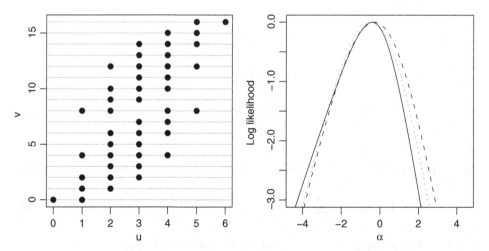

Figure 4.1 Conditional analysis of artificial data from a logistic regression model. Left: support of sufficient statistic (u, v). Right: log conditional likelihood $\ell_c(\alpha)$ (solid), profile log likelihood $\ell_p(\alpha)$ (dashes), and adjusted version $\ell_a(\alpha)$ (dots) when $(u, v) = (2, 4)$.

the possible values of (u, v). The density $f(v \mid u; \beta)$ is degenerate for $u = 0$ or $u = 6$, which correspond to all the responses equalling 0 or 1 respectively; otherwise the sets $\mathcal{Y}_u$ have at least five elements. The densities $f(u \mid v; \alpha)$ have at most four support points, but conditioning produces sensible inferences nevertheless. Suppose for example that the responses $y_1, \ldots, y_6$ equal $1, 0, 0, 0, 1, 0$, so that $(u, v) = (2, 4)$. The right-hand panel of the figure shows the log conditional likelihood

$$\ell_c(\alpha) = 2\alpha - \log\left(e^\alpha + \cdots + e^{4\alpha}\right),$$

the profile log likelihood $\ell_p(\alpha)$, and its adjusted version $\ell_a(\alpha)$. Adjustment shifts the profile log likelihood towards the log conditional likelihood, though ℓ_a and ℓ_c remain appreciably different; with just $n = 6$ binary observations this is not surprising; see Section 4.2.

To illustrate the smoothing effect of higher order approximation, we changed the covariate vector to $0.1, 1.2, 1.3, 2.4, 4.5, 8.6$. With this change there are 52 distinct values of v, so the conditional densities $f(u \mid v)$ are degenerate for almost every v, including that for the response values used above, for which $(u, v) = (2, 4.6)$. The closest values of v are 4.5 and 4.9, which also give degenerate conditional densities, so exact inference would yield a flat log likelihood in any of these cases. Nevertheless, the log profile and log adjusted likelihoods ℓ_p and ℓ_a exist and are close to those obtained using the original covariates. Thus higher order approximation involves an implicit smoothing step which can produce reasonable results even if the corresponding exact conditional density does not exist. This is particularly useful when there are many covariates and the appropriate subset of $\mathcal{Y}$ may be hard to find, even if it is not a singleton.

The difficulties with convergence of fitting algorithms may be illustrated by taking $y_1 = \cdots = y_4 = 1$, $y_5 = y_6 = 0$. Then, with the original covariates we have $(u, v) = (4, 4)$, and the estimates and their standard errors are $\widehat{\alpha} = 69$ (1.27×10^5) and $\widehat{\beta} = -23$ (4.2×10^4); the fitted probabilities then effectively equal the response values and an error is signalled by the R routine `glm`.

4.2 Urine data

Table 4.1 gives data concerning calcium oxalate crystals in samples of urine. The binary response r is an indicator of the presence of such crystals, and there are six explanatory variables: specific gravity (`gravity`), i.e. the density of urine relative to water; pH (`ph`); osmolarity (`osmo`, mOsm); conductivity (`conduct`, mMho); urea concentration (`urea`, millimoles per litre); and calcium concentration (`calc`, millimoles per litre). We drop the two incomplete cases 1 and 55, and the remaining 77 observations form the data frame `urine` of the `cond` package. A natural starting point for analysis is a logistic regression model with

$$\Pr(R_i = 1) = 1 - \Pr(R_i = 0) = \frac{\exp(x_i^\mathsf{T}\beta)}{1 + \exp(x_i^\mathsf{T}\beta)}, \quad i = 1, \ldots, n,$$

Table 4.1 *Data on the presence of calcium oxalate crystals in 79 samples of urine (Andrews and Herzberg, 1985, p. 249). See text for details.*

	r	gravity	ph	osmo	conduct	urea	calc
1	0	1.021	4.91	725	–	443	2.45
2	0	1.017	5.74	577	20.0	296	4.49
3	0	1.008	7.20	321	14.9	101	2.36
4	0	1.011	5.51	408	12.6	224	2.15
5	0	1.005	6.52	187	7.5	91	1.16
6	0	1.020	5.27	668	25.3	252	3.34
7	0	1.012	5.62	461	17.4	195	1.40
8	0	1.029	5.67	1107	35.9	550	8.48
9	0	1.015	5.41	543	21.9	170	1.16
10	0	1.021	6.13	779	25.7	382	2.21
11	0	1.011	6.19	345	11.5	152	1.93
12	0	1.025	5.53	907	28.4	448	1.27
13	0	1.006	7.12	242	11.3	64	1.03
14	0	1.007	5.35	283	9.9	147	1.47
15	0	1.011	5.21	450	17.9	161	1.53
16	0	1.018	4.90	684	26.1	284	5.09
17	0	1.007	6.63	253	8.4	133	1.05
18	0	1.025	6.81	947	32.6	395	2.03
19	0	1.008	6.88	395	26.1	95	7.68
20	0	1.014	6.14	565	23.6	214	1.45
21	0	1.024	6.30	874	29.9	380	5.16
22	0	1.019	5.47	760	33.8	199	0.81
23	0	1.014	7.38	577	30.1	87	1.32
24	0	1.020	5.96	631	11.2	422	1.55
25	0	1.023	5.68	749	29.0	239	1.52
26	0	1.017	6.76	455	8.8	270	0.77
27	0	1.017	7.61	527	25.8	75	2.17
28	0	1.010	6.61	225	9.8	72	0.17
29	0	1.008	5.87	241	5.1	159	0.83
30	0	1.020	5.44	781	29.0	349	3.04
31	0	1.017	7.92	680	25.3	282	1.06
32	0	1.019	5.98	579	15.5	297	3.93
33	0	1.017	6.56	559	15.8	317	5.38
34	0	1.008	5.94	256	8.1	130	3.53
35	0	1.023	5.85	970	38.0	362	4.54
36	0	1.020	5.66	702	23.6	330	3.98
37	0	1.008	6.40	341	14.6	125	1.02
38	0	1.020	6.35	704	24.5	260	3.46
39	0	1.009	6.37	325	12.2	97	1.19
40	0	1.018	6.18	694	23.3	311	5.64
41	0	1.021	5.33	815	26.0	385	2.66
42	0	1.009	5.64	386	17.7	104	1.22

43	0	1.015	6.79	541	20.9	187	2.64
44	0	1.010	5.97	343	13.4	126	2.31
45	0	1.020	5.68	876	35.8	308	4.49
46	1	1.021	5.94	774	27.9	325	6.96
47	1	1.024	5.77	698	19.5	354	13.00
48	1	1.024	5.60	866	29.5	360	5.54
49	1	1.021	5.53	775	31.2	302	6.19
50	1	1.024	5.36	853	27.6	364	7.31
51	1	1.026	5.16	822	26.0	301	14.34
52	1	1.013	5.86	531	21.4	197	4.74
53	1	1.010	6.27	371	11.2	188	2.50
54	1	1.011	7.01	443	21.4	124	1.27
55	1	1.022	6.21	–	20.6	398	4.18
56	1	1.011	6.13	364	10.9	159	3.10
57	1	1.031	5.73	874	17.4	516	3.01
58	1	1.020	7.94	567	19.7	212	6.81
59	1	1.040	6.28	838	14.3	486	8.28
60	1	1.021	5.56	658	23.6	224	2.33
61	1	1.025	5.71	854	27.0	385	7.18
62	1	1.026	6.19	956	27.6	473	5.67
63	1	1.034	5.24	1236	27.3	620	12.68
64	1	1.033	5.58	1032	29.1	430	8.94
65	1	1.015	5.98	487	14.8	198	3.16
66	1	1.013	5.58	516	20.8	184	3.30
67	1	1.014	5.90	456	17.8	164	6.99
68	1	1.012	6.75	251	5.1	141	0.65
69	1	1.025	6.90	945	33.6	396	4.18
70	1	1.026	6.29	833	22.2	457	4.45
71	1	1.028	4.76	312	12.4	10	0.27
72	1	1.027	5.40	840	24.5	395	7.64
73	1	1.018	5.14	703	29.0	272	6.63
74	1	1.022	5.09	736	19.8	418	8.53
75	1	1.025	7.90	721	23.6	301	9.04
76	1	1.017	4.81	410	13.3	195	0.58
77	1	1.024	5.40	803	21.8	394	7.82
78	1	1.016	6.81	594	21.4	255	12.2
79	1	1.015	6.03	416	12.8	178	9.39

where x_i represents the vector of explanatory variables associated with the ith observed response r_i, and β is the corresponding parameter vector. The resulting log likelihood is of linear exponential family form

$$\ell(\beta) = r^{\mathrm{T}} X \beta - \sum_{i=1}^{n} \log\left\{1 + \exp(x_i^{\mathrm{T}} \beta)\right\},$$

where the ith rows of the covariate matrix X and of the vector r are respectively x_i^T and r_i. Code 4.1 shows the R output obtained when this model is fitted to the data, with all six explanatory variables included; β is then a vector of length seven, including the intercept. The collinearity between `gravity` and the constant vector has been reduced by including $100(\texttt{gravity} - 1)$ rather than `gravity` itself. The only terms significant at the 5% level are `urea` and `calc`. As one might expect, an increase in the calcium concentration `calc` increases the probability that calcium crystals are present; the opposite is true for urea concentration `urea`.

The urine data have $n = 77$ observations and we fit a model with only seven parameters, so we might expect there to be little difference between first order and higher order inferences. We can check this using the `cond` package of the `hoa` package bundle. We take as two separate possible parameters of interest the coefficients for `urea` and `calc`. Figure 4.2 shows the profile log likelihoods and the approximate conditional log likelihoods for two conditional fits to the data, the R code for which is given in Code 4.1. This gives 95% confidence intervals for the coefficients of `urea` and `calc`, obtained from different first order and higher order statistics. The difference between the 95% confidence intervals based on the likelihood root and on the modified likelihood root is largest for `calc`, and as usual higher order correction moves the interval towards the origin; in fact the corrected interval for `urea` includes the origin. The nuisance parameter corrections r_{NP} from the decomposition (2.10) are both large, indicating that higher order solutions are preferable to first order ones.

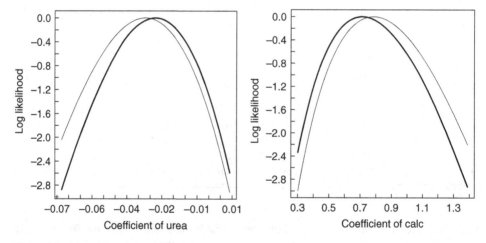

Figure 4.2 Comparison of log likelihoods for the urine data: profile log likelihood (solid line), approximate conditional log likelihood (bold line). The variables of interest are urea (left panel) and calcium concentration (right panel). The graphical output is obtained with the `plot` method of the `cond` package.

Code 4.1 Approximate conditional inference for the urine data.

```
> urine.glm <-
+    glm( formula = r ~ I(100*(gravity-1))+ph+osmo+conduct+urea+calc,
+         family = binomial, data = urine )

> summary( urine.glm )

Coefficients:
                       Estimate  Std. Error  z value  Pr(>|z|)
(Intercept)             0.60609     3.79582    0.160   0.87314
I(100 * (gravity - 1))  3.55944     2.22110    1.603   0.10903
ph                     -0.49570     0.56976   -0.870   0.38429
osmo                    0.01681     0.01782    0.944   0.34536
conduct                -0.43282     0.25123   -1.723   0.08493 .
urea                   -0.03201     0.01612   -1.986   0.04703 *
calc                    0.78369     0.24216    3.236   0.00121 **
---
Signif. codes:  0 *** 0.001 ** 0.01 * 0.05 . 0.1   1

    Null deviance:  105.17  on 76  degrees of freedom
Residual deviance:   57.56  on 70  degrees of freedom
AIC: 71.56

> urine.cond.urea <- cond( urine.glm, offset = urea )

> coef( urine.cond.urea )
          Estimate   Std. Error
uncond.   -0.03201      0.01612
cond.     -0.02759      0.01490

> summary( urine.cond.urea, coef = F )

Confidence intervals
--------------------
 level = 95 %
                                          lower two-sided         upper
Wald pivot                                     -0.06361      -0.0004208
Wald pivot (cond. MLE)                         -0.05679       0.0016100
Likelihood root                                -0.06677      -0.0024570
Modified likelihood root                       -0.05874       0.0004687
Modified likelihood root (cont. corr.)  -0.05889             0.0005680

Diagnostics:
-----------
     INF        NP
  0.05134   0.40329
```

Code 4.1 Approximate conditional inference for the urine data (cont.).

```
> urine.cond.calc <- cond( urine.glm, offset = calc )

> coef( urine.cond.calc )
          Estimate  Std.  Error
uncond.     0.7837        0.2422
cond.       0.7111        0.2283

> summary( urine.cond.calc, coef = F )

Confidence intervals
 level = 95 %
                                      lower two-sided    upper
Wald pivot                                   0.3091      1.258
Wald pivot (cond. MLE)                       0.2637      1.158
Likelihood root                              0.3815      1.342
Modified likelihood root                     0.3193      1.213
Modified likelihood root (cont. corr.)       0.3044      1.254

Diagnostics:
-----------
     INF        NP
0.08451   0.32878
```

A Bayesian analysis is also possible in this setting. Expression (2.16) shows that the higher order approximation to the marginal posterior distribution function uses

$$q_{\mathrm{B}}(\psi) = -\ell_{\mathrm{p}}'(\psi) j_{\mathrm{p}}(\widehat{\psi})^{-1/2} \frac{|j_{\lambda\lambda}(\psi, \widehat{\lambda}_{\psi})|^{1/2}}{|j_{\lambda\lambda}(\widehat{\psi}, \widehat{\lambda})|^{1/2}} \frac{\pi(\widehat{\psi}, \widehat{\lambda})}{\pi(\psi, \widehat{\lambda}_{\psi})},$$

where as usual ψ is the parameter of interest and λ is the nuisance parameter. Here we suppose that ψ is the component of β that corresponds to the covariate `calc` and use λ to denote the remaining components. In this model there is a natural non-informative prior, as discussed in Section 8.7; an expression for q with this choice of prior is given at (8.55). The quantity $\ell_{\mathrm{p}}'(\psi)$ may be computed as the derivative of a spline fitted to the profile log likelihood, and $j_{\mathrm{p}}(\widehat{\psi})^{-1/2}$ is the standard error for $\widehat{\psi}$; both of these may be obtained from the `workspace` component of the `urine.cond.calc` object. The remaining quantity needed is obtained from

$$j(\psi, \lambda) = X^{\mathrm{T}} W X, \quad W = \mathrm{diag}\{m_1 p_1(1-p_1), \ldots, m_n p_n(1-p_n)\},$$

where $p_i = p_i(\psi, \lambda)$ denotes the success probability for y_i and m_i the corresponding binomial denominator; in this case $m_i = 1$. It is straightforward to write a routine to evaluate q_{B} in this example; the analogous formulae for other linear exponential family models are equally simple. The 95% confidence interval for ψ computed using this Bayesian approach is $(0.3213, 1.211)$, the corresponding frequentist interval being $(0.3193, 1.213)$.

At first sight the difference between first and higher order approximations in this example is surprising. In order to elucidate it, note that one interpretation of a binary response is as a dichotomized version of an underlying continuous variable Z. If Z has the logistic distribution with mean $x^{\mathrm{T}}\beta$ and dispersion parameter $\sigma = 1$, then

$$\Pr(Z \leq z) = \frac{\exp(z - x^{\mathrm{T}}\beta)}{1 + \exp(z - x^{\mathrm{T}}\beta)}, \quad -\infty < z < \infty,$$

and so $R = 1$ if and only if $Z > 0$. To try to quantify the loss of information in reducing Z to R we created four larger sets of data by replicating each of the 77 original binary observations $m = 2, 4, 8$ and 16 times, and in each case computed the first and higher order approximations. Figure 4.3 shows how the functions $r(\psi)$ and $r^*(\psi)$ for the parameter ψ associated with the concentration of calcium depend on m. The key features are the increasing slope and straightness of both curves, and their increasing similarity as m

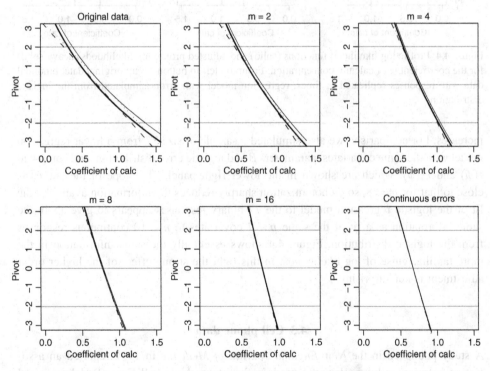

Figure 4.3 Pivot functions $r(\psi)$ (solid), $r^*(\psi)$ (heavy solid) and ordinary and conditional Wald statistics (dashed) for the coefficient ψ of calcium concentration. From top left to bottom right: original data, artificial binomial data obtained by replicating the original responses m times, and continuous data simulated from logistic distribution.

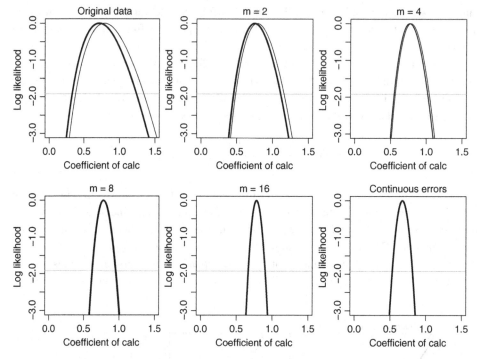

Figure 4.4 Profile log likelihood functions (solid) and adjusted profile log likelihoods (heavy solid) for the coefficient ψ of calcium concentration. From top left to bottom right: original data, binomial data with responses replicated m times, regression model with errors simulated from the logistic distribution.

increases. For comparison we also simulated a sample of size 77 from a linear regression model with the same covariates, parameters $\widehat{\beta}$ and logistic error distribution, and computed $r(\psi)$ and $r^*(\psi)$, which are shown in the lower right panel. The slope of the curves is close to that for $m = 8$, so dichotomization sharply reduces the information available: the fit of the logistic regression model to the 77 binary responses appears to give about the same information as a fit of the same $p = 7$ covariates to $n = 10$ continuous responses from the logistic distribution. Figure 4.4 shows essentially the same information in the more familiar guise of log likelihoods. In this light the strong effect of the higher order adjustment is not surprising.

4.3 Cell phone data

A study published in the *New England Journal of Medicine* in 1997 reported an association between cellular telephone usage and motor vehicle collisions (Redelmeier and Tibshirani, 1997a). The investigators collected data on 699 drivers who had cellular telephones and were involved in a motor vehicle collision resulting in substantial property damage but not personal injury. Each person's cellular telephone calls on the

Table 4.2 *Association between cellular telephone use and vehicle collisions (Redelmeier and Tibshirani, 1997a).*

		Control window	
		On telephone	Not on telephone
Hazard window	On telephone	13	157
	Not on telephone	24	505

day of the collision and during the previous week were analyzed through the use of detailed billing records. Table 4.2 shows a summary of the data in the form of a 2×2 table.

The basic analysis of the data was a comparison of telephone usage in a 10-minute *hazard window* before the crash with telephone usage in the same 10-minute *control window* on a preceding day. This form of case-control study in which each subject serves as his/her own control is sometimes called a case-crossover study. The investigators used a number of alternative control windows, such as the same 10-minute window one week earlier, on the previous weekday or weekend, depending on whether the case occurred on a weekday or weekend, and so forth. There is a bias in this comparison due to the fact that a driver may not have been driving the car during the control window, and hence not at risk for a collision. We ignore this source of bias here, but see Problem 25.

The data underlying Table 4.2 are in the form of matched pairs (y_{1i}, y_{2i}), for $i = 1, \ldots, n$, where

$$y_{1i} = \begin{cases} 1, & \text{call in hazard window,} \\ 0 & \text{otherwise,} \end{cases}$$

$$y_{2i} = \begin{cases} 1, & \text{call in control window,} \\ 0 & \text{otherwise.} \end{cases}$$

We will assume that y_{1i} and y_{2i} are realizations of independent Bernoulli variables Y_{1i} and Y_{2i} with success probabilities

$$p_{1i} = \Pr(Y_{1i} = 1) = \frac{e^{\psi + \beta_i}}{1 + e^{\psi + \beta_i}}, \quad p_{2i} = \Pr(Y_{2i} = 1) = \frac{e^{\beta_i}}{1 + e^{\beta_i}}.$$

If so, then the differences of log odds,

$$\log \left\{ \frac{\Pr(Y_{1i} = 1)}{\Pr(Y_{1i} = 0)} \right\} - \log \left\{ \frac{\Pr(Y_{2i} = 1)}{\Pr(Y_{2i} = 0)} \right\} = \psi + \beta_i - \beta_i = \psi,$$

are equal for each of the matched pairs. Under this model e^ψ is the retrospective odds ratio, that is,

$$\exp(\psi) = \frac{\Pr(\text{call} \mid \text{collision})/\Pr(\text{no call} \mid \text{collision})}{\Pr(\text{call} \mid \text{no collision})/\Pr(\text{no call} \mid \text{no collision})}$$

$$= \frac{p_{1i}}{1 - p_{1i}} \frac{1 - p_{2i}}{p_{2i}},$$

but as the probability of a collision is small, this approximates the odds ratio of interest, that is,

$$\exp(\psi) \approx \frac{\Pr(\text{collision} \mid \text{call})}{\Pr(\text{collision} \mid \text{no call})}.$$

The full likelihood based on $(Y_{11}, Y_{21}), \ldots, (Y_{1n}, Y_{2n})$ has $n + 1$ parameters, and the overall maximum likelihood estimate of ψ is not consistent as $n \longrightarrow \infty$ (Problem 26). As in logistic regression, however, the nuisance parameters β_i can be eliminated by conditioning. Here it is appropriate to condition on the totals for the pairs, giving

$$\Pr(Y_{1i} = 1 \mid Y_{1i} + Y_{2i} = 0) = 0, \tag{4.3}$$

$$\Pr(Y_{1i} = 1 \mid Y_{1i} + Y_{2i} = 1) = \frac{e^\psi}{1 + e^\psi},$$

$$\Pr(Y_{1i} = 1 \mid Y_{1i} + Y_{2i} = 2) = 1. \tag{4.4}$$

Pairs whose outcomes are equal have conditional probabilities (4.3) and (4.4) that do not depend on ψ, and so these pairs do not contribute to the conditional likelihood

$$\left(\frac{e^\psi}{1 + e^\psi}\right)^{n_{10}} \left(\frac{1}{1 + e^\psi}\right)^{n_{01}},$$

where n_{10} is the number of pairs $(y_{1i} = 1, y_{2i} = 0)$ and n_{01} is the number of pairs $(y_{1i} = 0, y_{2i} = 1)$. Writing $\gamma = e^\psi/(1 + e^\psi)$, the conditional likelihood based on the data in Table 4.2 is that for a binomial variable R with observed value 157, denominator $m = n_{01} + n_{10} = 181$ and success probability γ, that is,

$$L_c(\psi) = \binom{181}{157} \gamma^{157} (1 - \gamma)^{24}, \tag{4.5}$$

giving maximum conditional likelihood estimates $\widehat{\gamma}_c = 157/181 = 0.8674$ and $\widehat{\psi}_c = \log\{\widehat{\gamma}_c/(1 - \widehat{\gamma}_c)\} = 1.878$, and estimated odds ratio $\exp(\widehat{\psi}_c) = 6.54$; this last is the estimated risk of a collision while using the telephone relative to a collision occurring while not using the telephone. As mentioned above, this estimate is probably somewhat too large due to the possibility that the subject was not in fact driving in the control window, and a sub-study estimated the probability of this to be 0.35. The relative risk reported in the press was thus $4.25 = 6.54 \times 0.65$; see Problem 25. We can use the binomial distribution

Table 4.3 *Exact and approximate 95% confidence limits for* γ *and* e^{ψ} *based on the matched pairs analysis of the data in Table 4.2.* $\Phi(t_c)$ *and* $\Phi(r_c)$ *are first order approximations, and* $\Phi(r_c^*)$ *is a third order approximation.*

	Confidence interval for γ	Confidence interval for the relative risk e^{ψ}
Exact	(0.815, 0.913)	(4.419, 10.516)
$\Phi(t_c)$	(0.818, 0.917)	(4.494, 11.021)
$\Phi(r_c)$	(0.813, 0.912)	(4.346, 10.303)
$\Phi(r_c^*)$	(0.814, 0.913)	(4.390, 10.450)

of R to compute an exact confidence interval for γ, and hence e^{ψ}, by finding the smallest γ satisfying

$$\Pr(R > 157; \gamma_L) \geq \alpha/2$$

and the largest γ satisfying

$$\Pr(R \leq 157; \gamma_U) \geq \alpha/2.$$

The resulting confidence intervals are compared to first order and third order confidence intervals in Table 4.3. These were obtained using the log conditional likelihood derived from (4.5); in this exponential family model q is taken to be the Wald pivot t.

4.4 Multiple myeloma data

Table 4.4 contains data from a randomized trial comparing the effect of two chemotherapy treatments on survival of patients with multiple myeloma. The table contains outcomes for 156 patients at 21 institutions, an average of 7.4 patients per institution. As the responses are binary, the information on parameters in any reasonable model is likely to be small, and inference procedures should make allowance for this.

In applications like this interest sometimes focuses on the difference of risks between the two groups, which has a direct interpretation in terms of the percentage change in survival rates. We suppose that this difference is constant across the different institutions, and model the data as independent binomial variables: at the ith institution the numbers y_{1i} and y_{2i} of surviving patients who received the two treatments are binomial with denominators m_{1i} and m_{2i} and probabilities p_{1i} and p_{2i}. The risk difference for the ith institution is $p_{2i} - p_{1i}$, and we suppose that $p_{2i} - p_{1i} = \psi$ for $i = 1, \ldots, n$, where $n = 21$. The domain of the interest parameter ψ is the interval $-1 \leq \psi \leq 1$, so on writing

$$p_{1i} = \xi_i, \quad p_{2i} = \xi_i + \psi, \quad i = 1, \ldots, n,$$

we see that $\xi_- < \xi_i < \xi_+$ for all i, where

$$\xi_- \equiv \xi_-(\psi) = \max(0, -\psi), \quad \xi_+ \equiv \xi_+(\psi) = \min(1, 1 - \psi).$$

Table 4.4 *Data on survival of multiple myeloma patients following two different chemotherapy treatments in a Cancer and Leukaemia Group B trial (Lipsitz et al., 1998).*

Institution	y_1/m_1	y_2/m_2	Institution	y_1/m_1	y_2/m_2
1	3/4	1/3	12	2/2	0/2
2	3/4	8/11	13	1/4	1/5
3	2/2	2/3	14	2/3	2/4
4	2/2	2/2	15	2/4	4/6
5	2/2	0/3	16	4/12	3/9
6	1/3	2/3	17	1/2	2/3
7	2/2	2/3	18	3/3	1/4
8	1/5	4/4	19	1/4	2/3
9	2/2	2/3	20	0/3	0/2
10	0/2	2/3	21	2/4	1/5
11	3/3	3/3			

For numerical purposes it turns out to be better to write

$$\xi_i \equiv \xi(\lambda_i, \psi) = \xi_-(\psi) + \frac{\xi_+(\psi) - \xi_-(\psi)}{1 + \exp(-\lambda_i)}, \quad -\infty < \lambda_i < \infty,$$

so that the components of the full parameter vector are $\theta^T = (\psi, \lambda_1, \ldots, \lambda_n)$.

Although this model is an exponential family, it is not in canonical form, and exact elimination of the λ_i is impossible. However, following the discussion in Sections 2.4 and 8.5.3, we may use (2.11) with local parameter obtained using (2.15). It turns out that the first component of $\varphi(\theta)$ may be taken to be (Problem 27)

$$\sum_{i=1}^{n} m_{2i} \log \left\{ \frac{\psi + \xi(\lambda_i, \psi)}{1 - \psi - \xi(\lambda_i, \psi)} \right\},$$

and its remaining n components as

$$m_{2i} \log \left\{ \frac{\psi + \xi(\lambda_i, \psi)}{1 - \psi - \xi(\lambda_i, \psi)} \right\} + m_{1i} \log \left\{ \frac{\xi(\lambda_i, \psi)}{1 - \xi(\lambda_i, \psi)} \right\}, \quad i = 1, \ldots, n.$$

Once these computations have been performed, the numerical approach outlined in Section 2.4 yields the quantities needed for higher order inference. The R code for this is given in Code 4.2. Institutions 4, 11 and 20 must be dropped from the analysis, because as they have $y_{1i}/m_{1i} = y_{2i}/m_{2i} = 0$ or 1, a perfect fit can be obtained for any ψ, and they contribute no useful information to the likelihood.

Figure 4.5 shows how three approximate pivots vary as functions of the risk difference ψ. There is virtually no difference between the Wald pivot t and the likelihood root r, but a large correction shifts the 95% confidence interval of $(-0.081, 0.243)$ based on r

Code 4.2 R code for analysis of risk difference for the multiple myeloma data.

```
n1 <- c(4, 4,2,2,2,3,2,5,2,2,3,2,4,3,4,12,2,3,4,3,4)
y1 <- c(3, 3,2,2,2,1,2,1,2,0,3,2,1,2,2, 4,1,3,1,0,2)
n2 <- c(3,11,3,2,3,3,3,4,3,3,3,2,5,4,6, 9,3,4,3,2,5)
y2 <- c(1, 8,2,2,0,2,2,4,2,2,3,0,1,2,4, 3,2,1,2,0,1)

risk <- data.frame( n1, y1, n2, y2 )

nlogL <- function(psi, lam, d)
{
 xi0 <- max(0, -psi)
 xi1 <- min(1, 1 - psi)
 xi <- xi0 + (xi1-xi0) / (1 + exp(-lam))
 p1 <- psi + xi
 p2 <- xi
 -sum( d$y1*log(p1) + (d$n1-d$y1) * log(1-p1) +
         d$y2*log(p2) + (d$n2-d$y2) * log(1-p2) )
}

make.V <- function(th, d)  NULL

phi <- function(th, V, d)
{
 n <- nrow(d)
 out <- rep(0, n + 1)
 psi <- th[1]
 lam <- th[-1]
 xi0 <- max(0, -psi)
 xi1 <- min(1, 1 - psi)
 xi <- xi0 + (xi1-xi0)/(1 + exp(-lam))
 p1 <- psi + xi
 p2 <- xi
 out1 <- d$n1 * log( p1/(1-p1) )
 out2 <- d$n2 * log( p2/(1-p2) )
 c( sum(out1), apply( cbind(out1, out2), 1, sum ) )
}

bad <- c (4, 11, 20)
psi.val <- seq(from = -0.2, to = 0.3, length = 50)
risk.init <- rep(0, 22)

risk.fr <-
   fraser.reid( psi = psi.val, nlogL = nlogL, phi = phi, make.V = make.V,
              th.init = risk.init[-bad], d = risk[-bad,] )
```

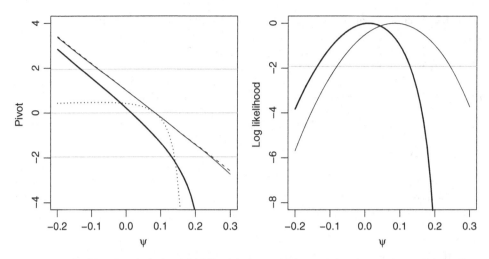

Figure 4.5 Likelihood analysis for risk difference for multiple myeloma data. Left panel: behaviour as functions of ψ of three approximate pivots, the likelihood root $r(\psi)$ (solid), the modified likelihood root $r^*(\psi)$ (heavy), and the Wald pivot $t(\psi)$ (dashes). Also shown is $q(\psi)$ (dots). The horizontal lines are at $0, \pm 1.96$. Right panel: profile log likelihood (solid) and adjusted profile log likelihood (heavy). The horizontal line at -1.92 shows limits of 95% confidence intervals based on the χ_1^2 approximation for the likelihood ratio statistic.

to an interval of $(-0.128, 0.139)$ based on r^*. This correction changes the tail probability for testing $\psi = 0$ against the hypothesis $\psi > 0$ from 0.16, marginally suggestive of a difference in risks, to 0.45, which gives no evidence whatever of such a difference.

The right-hand panel of the figure shows a strong difference between the profile and adjusted profile log likelihoods, which corresponds to the sharp difference between the behaviour of r and r^* as functions of ψ.

4.5 Speed limit data

In order to assess the effect on the accident rate on motorways when a speed limit was imposed, an experiment was conducted in Sweden on $n = 92$ days of the years 1961 and 1962, matched so that day i in 1961 was comparable to day i in 1962. On some days the speed limit was in effect and enforced, and not on other days. The number of traffic accidents with personal injuries reported to the police is given in Table 4.5.

A natural model for these data is that the numbers of accidents on day i in 1961 and in 1962, which we denote by Y_{1i} and Y_{2i}, are independent Poisson variables with means μ_{1i} and μ_{2i}, where

$$\mu_{1i} = \exp(\lambda_i + \psi x_{1i}), \quad \mu_{2i} = \exp(\lambda_i + \delta + \psi x_{2i}), \quad i = 1, \ldots, n.$$

In this log-linear model $\lambda_1, \ldots, \lambda_n$ represent background accident rates for the different days, δ represents a possible overall change in the accident rate from 1961 to 1962, and

Table 4.5 *Numbers of accidents* y_1 *and* y_2 *occurring on Swedish motorways on 92 matched days in 1961 and 1962; a + indicates that a speed limit was in place and enforced (Svensson, 1981).*

	y_1	y_2		y_1	y_2		y_1	y_2		y_1	y_2
1	9	9	24	12+	19+	47	37	29+	70	17	21
2	11	20	25	41+	32+	48	32	17+	71	16	20
3	9	15	26	15+	22+	49	25	17+	72	19	19
4	20	14	27	18+	24+	50	20	15+	73	18	20
5	31	30	28	11+	9+	51	40	25+	74	22	29
6	26	23	29	19+	10	52	21	9+	75	37	48
7	18	15	30	19+	14	53	18	16+	76	29	36
8	19	14	31	9+	18	54	35	25+	77	18	15
9	18	16	32	21+	26	55	21	25+	78	14	16
10	13	20	33	22+	38	56	25	16+	79	14	29
11	29	17+	34	23+	31	57	34	22+	80	18	12
12	40	23+	35	14+	12	58	42	21+	81	21	24+
13	28	16+	36	19+	8	59	27	17+	82	39	26+
14	17	20+	37	15+	22	60	34	26+	83	39	16+
15	15	13+	38	13+	17	61	47	41+	84	21	15+
16	21	13+	39	22+	31	62	36	25+	85	15	12+
17	24	9+	40	42+	49	63	15	12+	86	17	22+
18	15	10+	41	29+	23	64	26	17+	87	20	24+
19	32	17+	42	21+	14	65	27	21	88	24	16+
20	22	12+	43	12+	25	66	18	19	89	30	25+
21	24	7+	44	16+	24	67	16	24	90	25	14+
22	11	11+	45	17	18	68	32	44	91	8	15+
23	27	15+	46	27	19	69	28	31	92	21	9+

ψ represents the effect of imposing a speed limit. The covariates x_{1i} and x_{2i} are indicator variables taking value one if a speed limit was imposed on day i in 1961 and 1962 respectively, and taking value zero otherwise. In this case the conditional distribution of Y_{2i} given the value s_i of the total $S_i = Y_{1i} + Y_{2i}$ has a binomial distribution with denominator s_i and probability

$$\frac{\mu_{2i}}{\mu_{1i} + \mu_{2i}} = \frac{\exp(\lambda_i + \delta + \psi x_{2i})}{\exp(\lambda_i + \psi x_{1i}) + \exp(\lambda_i + \delta + \psi x_{2i})}$$
$$= \frac{\exp\{\delta + \psi(x_{2i} - x_{1i})\}}{1 + \exp\{\delta + \psi(x_{2i} - x_{1i})\}}.$$

Thus, conditional on the total number s_i of accidents on day i, the number of accidents y_{2i} in 1962 follows a logistic regression model with single covariate $z_i = x_{2i} - x_{1i}$, in which the nuisance parameters $\lambda_1, \ldots, \lambda_n$ do not appear.

The totals $S_i = Y_{1i} + Y_{2i}$ are independent Poisson variables with means

$$\xi_i = e^{\lambda_i} \left(e^{\psi x_{1i}} + e^{\delta + \psi x_{2i}} \right),$$

and only these means are identifiable from the S_i. Thus the likelihood contribution from the data y_{1i}, y_{2i} on the ith day may be factorized as

$$f(y_{1i}, y_{2i}; \lambda_i, \delta, \psi) = f(y_{2i} \mid s_i; \delta, \psi) f(s_i; \xi_i),$$

showing that no information about δ and ψ is lost by restricting inference to the binomial distributions of Y_{2i} given $S_i = s_i$. This is stronger than the usual situation with linear exponential families, as the parameters separate completely in the conditional and marginal components. Thus the conditional likelihood from which $\lambda_1, \ldots, \lambda_n$ have been eliminated equals the profile likelihood for the interest parameters ψ, δ, and likelihood inferences about the interest parameters – such as profile likelihood plots, maximum likelihood estimates, confidence intervals and test statistics – are identical under the Poisson and binomial setups. The logistic regression has the minor computational advantage of avoiding fitting a model with 94 parameters to 184 observations, so the residuals from the binomial model are much more nearly independent than are those from the log-linear model, but there is no advantage to the binomial model in terms of inference on the parameters of interest. The likelihood ratio statistics for successive inclusion of z_i and the year effect equal 59.3 and 0.7, each with one degree of freedom. Thus there is overwhelmingly strong evidence of an effect of the speed limit, but none of a difference between the years, once the limit has been included. The corresponding parameter estimates are -0.292 (0.043) and -0.029 (0.035), so the effect of a limit is to reduce the accident rate by a factor $\exp(-0.29) = 0.75$, with 0.95 confidence interval $\exp(-0.29 \pm 1.96 \times 0.043) = (0.69, 0.81)$.

In this case there is no discernible difference between confidence intervals based on first order and third order asymptotics.

An alternative approach to dealing with the nuisance parameters $\lambda_1, \ldots, \lambda_n$ is to assume that they are drawn randomly from a distribution $g(\lambda)$, and to integrate them out of the likelihood. Under this random effects model the likelihood contribution from the data for the ith day is

$$\int f(y_{1i}, y_{2i}; \lambda_i, \delta, \psi) g(\lambda_i) \, d\lambda_i, \quad i = 1, \ldots, n, \qquad (4.6)$$

and the full likelihood is the product of these terms, which depend on δ, ψ and g, but not on the λ_i. Inference when such integrals appear in the likelihood is a topic of much current research interest, and approximation to the integrals has been advocated using variants of Laplace approximation, Markov chain Monte Carlo and the EM algorithm. Here, however, the parameters of interest correspond to contrasts within values of i, and so the use of random effects modelling for day effects is irrelevant, as (4.6) may be written as

$$f(y_{2i} \mid s_i; \delta, \psi) \int f(s_i; \lambda_i, \delta, \psi) g(\lambda_i) \, d\lambda_i, \quad i = 1, \ldots, n,$$

and as before inference may be based on the first, binomial, terms. There is a close analogy with the analysis of a split-plot experiment, where plot level variation does not affect contrasts of treatments within plots. Although information about ψ and δ is in principle retrievable from the integral, it comes at the expense of specifying the density g of the random effects, and in most cases would come at too high a price.

4.6 Smoking data

Table 4.6 summarises data from a classical epidemiological study on the relation between smoking and lung cancer in British male physicians. It shows the man-years at risk and the number of individuals dying of lung cancer, cross-classified by the number of cigarettes smoked daily and the number of years of smoking, taken to be age minus twenty years. The man-years at risk in each cell, T, is the total period for which the individuals in that cell were at risk of death.

Death from lung cancer is a rare event, and one possible model is that the number of deaths in a cell of the table has a Poisson distribution with mean

$$\mu(\theta) = T \times e^{\theta_1} d^{\theta_2} \left(1 + e^{\theta_3} c^{\theta_4}\right), \quad -\infty < \theta_1, \theta_3 < \infty, \quad \theta_2, \theta_4 > 0. \quad (4.7)$$

This has four unknown parameters, and presupposes power-law dependence of death rate on exposure duration, d, and cigarette consumption, c. The background death-rate in the absence of smoking, the death-rate for non-smokers, is $e^{\theta_1} d^{\theta_2}$, which represents the overall effect of other causes of lung cancer. In the computations below we take c and d to be the mid-points of the consumptions and durations given in Table 4.6, with $c = 40$ for the 35+ category, and we replace T by $T \times 10^{-5}$.

The log likelihood for a Poisson response model can be written as

$$\ell(\theta) = \sum_{i=1}^{n} \{y_i \log \mu_i(\theta) - \mu_i(\theta)\},$$

where y_i is the number of deaths in the ith cell of the table and $\mu_i(\theta)$ is its mean. The maximum likelihood estimates and their standard errors based on observed information are $\widehat{\theta}_1 = -13.77 \ (1.37)$, $\widehat{\theta}_2 = 4.469 \ (0.329)$, $\widehat{\theta}_3 = -1.374 \ (1.095)$ and $\widehat{\theta}_4 = 1.326 \ (0.223)$.

One aspect of potential interest is the value of θ_4, as $\theta_4 = 1$ would correspond to linear increase in death rate with c. The value of the likelihood root r corresponding to $\theta_4 = 1$ is 1.585, so the one-sided significance level for testing linear dependence of the death rate on c is 0.057.

Higher order inference entails finding the local parametrization φ given at (2.12). The log likelihood may be written in the form (2.13) with $\alpha_i(\theta) = \log \mu_i(\theta)$, so

$$\varphi(\theta) = \sum_{i=1}^{n} \left. \frac{\partial \mu_i(\theta)}{\partial \theta} \right|_{\theta = \widehat{\theta}} \log \mu_i(\theta).$$

Table 4.6 *Lung cancer deaths in British male physicians (Frome, 1983). The table gives man-years at risk/number of cases of lung cancer, cross-classified by years of smoking, taken to be age minus 20 years, and number of cigarettes smoked per day. For non-smokers, duration of smoking is age group minus 20 years.*

Duration of smoking d (years)	Daily cigarette consumption c						
	Nonsmokers	1–9	10–14	15–19	20–24	25–34	35+
5–19	10 366/1	3121	3577	4317	5683	3042	670
20–24	8162	2937	3286/1	4214	6385/1	4050/1	1166
25–29	5969	2288	2546/1	3185	5483/1	4290/4	1482
30–34	4496	2015	2219/2	2560/4	4687/6	4268/9	1580/4
35–39	3512	1648/1	1826	1893	3646/5	3529/9	1336/6
40–44	2201	1310/2	1386/1	1334/2	2411/12	2424/11	924/10
45–49	1421	927	988/2	849/2	1567/9	1409/10	556/7
50–54	1121	710/3	684/4	470/2	857/7	663/5	255/4
55–59	826/2	606	449/3	280/5	416/7	284/3	104/1

Table 4.7 *First order and higher order 95% confidence intervals for parameters of model for smoking data.*

	Wald pivot t	Likelihood root r	Modified likelihood root r^*
θ_1	$(-16.46, -11.09)$	$(-16.57, -11.18)$	$(-16.36, -11.04)$
θ_2	$(3.824, 5.115)$	$(3.838, 5.132)$	$(3.836, 5.129)$
θ_3	$(-3.521, 0.771)$	$(-3.746, 0.703)$	$(-3.861, 0.587)$
θ_4	$(0.888, 1.763)$	$(0.929, 1.834)$	$(0.925, 1.842)$

The model (4.7) above gives

$$\frac{\partial \mu_i(\theta)}{\partial \theta} = T_i e^{\theta_1} d_i^{\theta_2} \left(1 + e^{\theta_3} c_i^{\theta_4}, (1 + e^{\theta_3} c_i^{\theta_4}) \log d_i, e^{\theta_3} c_i^{\theta_4}, e^{\theta_3} c_i^{\theta_4} \log c_i \right)^{\mathrm{T}},$$

where terms involving consumption c are understood to vanish for non-smokers. The numerical approach sketched in Section 2.4 may then be used to obtain the various components of (8.34). When $\theta_4 = 1$, the result is $q_2 = 1.501$, so $r^* = 1.564$, giving a significance level of 0.059. This small change from the previous level is in the direction away from significance. Graphs of the pivots corresponding to r and r^* show that the largest change is for large values of θ_4, but it is not dramatic.

Table 4.7 compares first order and higher order confidence intervals for all the parameters of this model. Despite the generally low counts in Table 4.6, the effect of higher order correction is smaller than the differences between the Wald pivot and the likelihood root, except for θ_3.

Bibliographic notes

The argument for conditional inference in linear exponential families could be regarded as one of convenience, but the classical theory of similar hypothesis tests implies that inference for the parameter of interest without regard to the value of the nuisance parameter is only possible through conditioning. See for example Cox and Hinkley (1974, Chapter 5). Computation of exact conditional densities such as (4.2) and significance levels derived therefrom is available using specialised algorithms, which are available in software packages such as *StatXact* and *LogXact* (www.cytel.com), although the approximations we discuss have the advantage of implementing some smoothing, as discussed in Section 4.1. Conditional inference in generalized linear models is discussed by McCullagh and Nelder (1989, Chapter 11).

The cell phone study was reported in Redelmeier and Tibshirani (1997a) and has been widely cited since. The statistical aspects are described in more detail in Tibshirani and Redelmeier (1997), and a non-technical statistical discussion is given in Redelmeier and Tibshirani (1997b).

The factorisation of the joint density into conditional and marginal components is extensively discussed in Barndorff-Nielsen (1978); see also Jørgensen (1993). In the Poisson or log-linear model, the factorisation completely separates the parameters; this is referred to as a *cut* in the parameter space. See also Davison (2003, Section 5.2).

The practical implementation of conditional inference in logistic regression using standard likelihood pivots was discussed by Davison (1988) and substantially extended by Brazzale (2000). The 'crying babies' data from Cox (1970) and used by those authors is available in Cox and Snell (1981, Set 1), and is included in the cond package. Cox and Snell (1981) give several other examples of regression with binary and count data.

5

Regression with continuous responses

5.1 Introduction

In this chapter we consider models for data with continuous response values. The most common such model, the linear regression model with normal errors, is a very widely used statistical tool, and its variants and extensions are legion. Below we discuss both linear and nonlinear regression, with normal and non-normal errors.

In a linear regression model the response variables $y_1, \ldots, y_n$ are related to explanatory variables x_i by

$$y_i = x_i^{\mathsf{T}} \beta + \sigma \varepsilon_i, \quad i = 1, \ldots, n, \tag{5.1}$$

where x_i is a known $p \times 1$ vector, and the unknown parameters are the $p \times 1$ vector β and the positive scalar σ. We assume the random variables ε_i are independent and generated from a known continuous density function f. If this is the standard normal density then y_i is normally distributed with mean $x_i^{\mathsf{T}} \beta$ and variance σ^2. Exact confidence intervals and tests based on the t and F distributions may then be computed using the least squares estimates provided by the `lm` routine in R, or by equivalent procedures in other statistical packages. These estimates are notoriously sensitive to outliers, however, and it may be desired to use a longer-tailed distribution for the responses.

For any continuous error distribution, exact inference for the parameters may be obtained from the distribution of an appropriate pivot, conditional on an ancillary statistic known as the *configuration*, which is defined in terms of the maximum likelihood estimators $\widehat{\beta}$ and $\widehat{\sigma}$ as

$$a = (a_1, \ldots, a_n), \quad \text{where } a_i = (y_i - x_i^{\mathsf{T}} \widehat{\beta}) / \widehat{\sigma}.$$

There is a one-to-one transformation from $(y_1, \ldots, y_n)$ to the vector $(\widehat{\beta}, \widehat{\sigma}, a)$, and the joint density of the latter factors as

$$f_1(\widehat{\beta}, \widehat{\sigma} \mid a; \beta, \sigma) f_2(a); \tag{5.2}$$

this extends the result discussed in Section 8.5.2 for the simple location model. The joint conditional density of $(\widehat{\beta}, \widehat{\sigma})$ may be obtained exactly by renormalizing the likelihood function.

Inference for a single component of β, say β_1, is then based on the marginal distribution of $(\widehat{\beta}_1 - \beta_1)/v_1^{1/2}$, where v_1 is the estimated variance of the maximum likelihood estimator $\widehat{\beta}_1$. This marginal distribution eliminates the nuisance parameters $\beta_2, \ldots, \beta_p$ and σ. Similarly inference for σ may be performed using the marginal distribution of the pivot $\widehat{\sigma}/\sigma$ or equivalently that of $\log\widehat{\sigma} - \log\sigma$. We refer below to the 'marginal' densities or distributions for inference based on pivots such as $(\widehat{\beta}_1 - \beta_1)/v_1^{1/2}$ and $\widehat{\sigma}/\sigma$, but these densities are first conditional on a, as at (5.2).

Exact computation of the marginal distributions and densities entails p-dimensional numerical integration to eliminate the nuisance parameters. Higher order approximations can be obtained through the theoretical development outlined in Section 8.6.2. It can be shown that the confidence intervals obtained from this approach are identical to Bayesian posterior credible intervals, using the improper prior density $\pi(\beta, \sigma)\mathrm{d}\beta\,\mathrm{d}\sigma \propto \mathrm{d}\beta\,\mathrm{d}\sigma/\sigma$; see the bibliographic notes. The corresponding marginal posterior density functions yield marginal likelihoods for the parameters.

In the case of normal errors, the marginal density for inference about σ is that of the residual sum of squares RSS, which has the $\sigma^2\chi^2_{n-p}$ distribution when p explanatory variables are fitted, provided the model is correct. The maximum marginal likelihood estimator is the unbiased quantity $\mathrm{RSS}/(n-p)$, which stands in contrast to the downwardly-biased maximum likelihood estimator RSS/n obtained by maximizing the full log likelihood.

In Section 5.2 we use Student t_ν errors with model (5.1) and compare the results will those obtained when assuming normal errors, for various choices of the degrees of freedom ν.

A useful extension of model (5.1) is to replace the linear predictor $x_i^{\mathsf{T}}\beta$ by a known nonlinear function $\mu(x_i; \beta)$, called the mean function, giving

$$y_i = \mu(x_i; \beta) + \sigma\varepsilon_i, \quad i = 1, \ldots, n, \tag{5.3}$$

where x_i is typically a scalar and as before β is an unknown p-dimensional parameter. If the ε_is are normally distributed with mean zero and unit variance, model (5.3) is usually called the nonlinear regression model. Such models are widely used, especially for dose–response curves in bioassays. They can be fitted by nonlinear least squares or equivalently by maximum likelihood, for example using the `nls` routine in R. In the examples of this chapter we use the `nlreg` package in the `hoa` bundle, which allows variance heterogeneity and provides residual and diagnostic measures. Regression diagnostics are very important for assessing the adequacy of the fit, but detailed discussion of them is outside the scope of this book. Examples are given in Section 5.3 and in the herbicide case study of Section 6.4.

Inference on the regression coefficients and the variance parameter σ is commonly based on first order asymptotics. However, if the mean function is highly nonlinear, first order approximations may prove inaccurate. This is especially true of the large sample normal approximation to the distribution of the Wald pivot, which does not properly account for the nonlinearity of the model. Higher order inference for (5.3) is used for the *Daphnia magna* data example in Section 5.3. Other very useful tools for nonlinear regression are graphical summaries of the fit, such as profile plots, profile traces and contour plots. Profile plots are plots of the appropriate pivot as a function of a scalar parameter of interest, and were used in many of the examples in earlier chapters. Profile traces and contour plots are built into the `nlreg` package, and allow some assessment of the dependence between pairs of parameters.

In settings where (5.3) describes the mean function and there are groups of replicate observations at the design points x_i, the resulting information on how the variance changes makes it possible to model the variance as well as the mean. A more general form of model (5.3) is

$$y_{ij} = \mu(x_i; \beta) + \sigma_i \varepsilon_{ij}, \quad i = 1, \ldots, m, \quad j = 1, \ldots, n_i, \tag{5.4}$$

where m is the number of design points, n_i the number of replicates at design point x_i, y_{ij} represents the response of the jth experimental unit at the ith design point, and the errors ε_{ij} are independent $N(0, 1)$ variates. The definition of the model is completed by assuming that $\sigma_i^2 = \sigma^2 w(x_i; \beta, \rho)^2$, where σ^2 and ρ are respectively scalar and q-dimensional parameters. If $w(\cdot)^2$ is constant, we obtain the homoscedastic nonlinear regression model (5.3).

The log likelihood function corresponding to model (5.4) is

$$\ell(\beta, \rho) = -\sum_{i=1}^{m} \frac{n_i}{2} \log \sigma^2 - \sum_{i=1}^{m} \frac{n_i}{2} \log w(x_i; \beta, \rho)^2 - \sum_{i=1}^{m} \sum_{j=1}^{n_i} \frac{\{y_{ij} - \mu(x_i; \beta)\}^2}{2\sigma^2 w(x_i; \beta, \rho)^2}.$$

The maximum likelihood estimate of σ^2 is available in closed form and equals the average residual sum of squares RSS$/n$. Maximization with respect to β and ρ is usually accomplished via a two-step iterative procedure, where one maximizes alternately with respect to β and to ρ.

As with the constant variance case, inference on the regression coefficients and the variance parameters is typically based on first order results and linearization techniques, and profile plots, profile traces and contour plots are used as graphical summaries of the model fit. In addition to nonlinearity of the mean function, variance heterogeneity can also lead to substantial inaccuracies if the sample size is small or moderate. The estimators of β and of σ^2 and ρ are asymptotically correlated, unless the variance function is of the form $\sigma^2 w(x; \rho)^2$ and thus does not depend on the regression coefficients β. This problem is especially acute when the parameter of interest is a variance parameter, and then a substantial bias may arise unless the inference procedure accounts for the estimation of β. Higher order inference for nonlinear regression with heteroscedastic errors is illustrated on the radioimmunoassay data in Section 5.4, using procedures outlined in Section 8.6.3.

In Sections 5.5 and 5.6 we consider the analysis of survival time or failure time data. Parametric models that have proved useful for the analysis of such data include the gamma, Weibull, log-normal and log-logistic models, which are all of the form (5.1) where $y_i = \log t_i$ is the log failure time. Most survival data include some data that are censored, either because the study ended or the subject or item under investigation was removed from the study. Under so-called Type II censoring, the number of failures is fixed in advance, and testing is halted when that number of failures has occurred. The data then take the form $(y_{(1)}, y_{(2)}, \ldots, y_{(r)}, y^+_{(r)}, \ldots, y^+_{(r)})$, where the $y_{(i)}$ are ordered failure times, and the last $n - r$ observations are censored at $y_{(r)}$. This type of censoring often arises in life testing applications in reliability. In medical applications random censoring or time censoring is more common. Under Type II censoring the regression-scale model still admits a factorization of the form (5.2), and exact or approximate conditional inference for component parameters is obtained by a slight modification of the results for (5.1).

5.2 Nuclear power station data

Table 5.1 contains data on the construction of $n = 32$ light water reactors used in Cox and Snell (1981, Example G) to illustrate the application of linear regression modelling. The response variable is the cost of construction of a reactor (in US dollars$\times 10^{-6}$ adjusted to a 1976 base), and the other variables are taken to be explanatory. Costs are typically relative, so we take `log(cost)` as the response y. We also take logs of the other quantitative covariates, fitting linear models using `date`, `log(T1)`, `log(T2)`, `log(cap)`, `PR`, `NE`, `CT`, `log(N)`, `BW` and `PT`. The last of these indicates six plants for which there were partial turnkey guarantees, and some manufacturers' subsidies may be hidden in these costs.

There is a large literature on the selection of models in regression, a review of which is outside the scope of this book. For illustration we used the variable selection criterion AIC and its small sample version AIC_c (Problem 33), given by

$$\text{AIC} = n \log \text{RSS} + 2p, \quad \text{AIC}_c = n \log \text{RSS} + \frac{n(n+p)}{n-p-2},$$

where RSS is the residual sum of squares corresponding to the fitted model, and p the number of its regression parameters. When a linear regression model with normal errors is fitted to the data, the model which minimises both AIC and AIC_c has the explanatory variables shown in Table 5.2; these are also the variables found by backward elimination starting from the full model. Most of these variables are highly significant, but the variables `log(N)` and `PT`, representing the log cumulative number of stations previously built by the architect and the partial turnkey guarantee, are only borderline significant. A probability plot of the residuals suggests that the response distribution may be slightly more peaked than is the normal distribution, and it seems worthwhile to investigate to what extent the estimates and resulting conclusions depend on the normality assumption.

In light of the residual plot we next fit a model in which the responses have Student t_ν distributions with $\nu = 4$ degrees of freedom. Higher order inference for such models is discussed in Section 8.6.2, and they can be fitted using the code shown in Code 5.1, which uses the `marg` package in the `hoa` bundle. Table 5.2 shows the estimates and standard errors for this model, using both first order and higher order asymptotics. The first

Table 5.1 *Data on light water reactors (LWR) constructed in the USA (Cox and Snell, 1981, p. 81). The covariates are* date *(date construction permit issued),* T1 *(time between application for and issue of permit),* T2 *(time between issue of operating license and construction permit),* cap *(power plant capacity in MWe),* PR *(=1 if LWR already present on site),* NE *(=1 if constructed in north-east region of USA),* CT *(=1 if cooling tower used),* BW *(=1 if nuclear steam supply system manufactured by Babcock–Wilcox),* N *(cumulative number of power plants constructed by each architect-engineer),* PT *(=1 if partial turnkey plant).*

	cost	date	T1	T2	cap	PR	NE	CT	BW	N	PT
1	460.05	68.58	14	46	687	0	1	0	0	14	0
2	452.99	67.33	10	73	1065	0	0	1	0	1	0
3	443.22	67.33	10	85	1065	1	0	1	0	1	0
4	652.32	68.00	11	67	1065	0	1	1	0	12	0
5	642.23	68.00	11	78	1065	1	1	1	0	12	0
6	345.39	67.92	13	51	514	0	1	1	0	3	0
7	272.37	68.17	12	50	822	0	0	0	0	5	0
8	317.21	68.42	14	59	457	0	0	0	0	1	0
9	457.12	68.42	15	55	822	1	0	0	0	5	0
10	690.19	68.33	12	71	792	0	1	1	1	2	0
11	350.63	68.58	12	64	560	0	0	0	0	3	0
12	402.59	68.75	13	47	790	0	1	0	0	6	0
13	412.18	68.42	15	62	530	0	0	1	0	2	0
14	495.58	68.92	17	52	1050	0	0	0	0	7	0
15	394.36	68.92	13	65	850	0	0	0	1	16	0
16	423.32	68.42	11	67	778	0	0	0	0	3	0
17	712.27	69.50	18	60	845	0	1	0	0	17	0
18	289.66	68.42	15	76	530	1	0	1	0	2	0
19	881.24	69.17	15	67	1090	0	0	0	0	1	0
20	490.88	68.92	16	59	1050	1	0	0	0	8	0
21	567.79	68.75	11	70	913	0	0	1	1	15	0
22	665.99	70.92	22	57	828	1	1	0	0	20	0
23	621.45	69.67	16	59	786	0	0	1	0	18	0
24	608.80	70.08	19	58	821	1	0	0	0	3	0
25	473.64	70.42	19	44	538	0	0	1	0	19	0
26	697.14	71.08	20	57	1130	0	0	1	0	21	0
27	207.51	67.25	13	63	745	0	0	0	0	8	1
28	288.48	67.17	9	48	821	0	0	1	0	7	1
29	284.88	67.83	12	63	886	0	0	0	1	11	1
30	280.36	67.83	12	71	886	1	0	0	1	11	1
31	217.38	67.25	13	72	745	1	0	0	0	8	1
32	270.71	67.83	7	80	886	1	0	0	1	11	1

Table 5.2 *Parameter estimates and standard errors for linear models fitted to nuclear power station data. Shown are the results of a fit with normal responses, and for a fit with t_4 responses, both with first order asymptotics, for which are given the maximum likelihood estimates and standard errors based on the inverse observed information matrix, and with higher order asymptotics, for which are given the maximum likelihood estimates and standard errors based on the marginal likelihood for each parameter. In each case z is the ratio of the estimate to its standard error.*

| | Normal | | t_4, first order | | t_4, higher order | |
	Est (SE)	z	Est (SE)	z	Est (SE)	z
Constant	−13.26 (3.140)	−4.22	−11.30 (3.67)	−3.01	−11.86 (3.70)	−3.21
date	0.212 (0.043)	4.91	0.191 (0.048)	3.97	0.196 (0.049)	4.02
log(cap)	0.723 (0.119)	6.09	0.648 (0.113)	5.71	0.682 (0.129)	5.31
NE	0.249 (0.074)	3.36	0.242 (0.077)	3.12	0.239 (0.080)	2.97
CT	0.140 (0.060)	2.32	0.144 (0.054)	2.68	0.143 (0.063)	2.26
log(N)	−0.088 (0.042)	−2.11	−0.060 (0.043)	−1.40	−0.072 (0.048)	−1.51
PT	−0.226 (0.114)	−1.99	−0.282 (0.101)	−2.80	−0.265 (0.110)	−2.42

Code 5.1 R code for analysis of nuclear power station data.

```
library(marg)
data(nuclear)

# Fit normal-theory linear model and examine its contents

nuc.norm <- lm( log(cost) ~ date + log(cap) + NE + CT + log(N) + PT,
                data = nuclear)
summary(nuc.norm)

# Fit linear model with t errors and 4 df and examine its contents

nuc.t4 <- rsm( log(cost) ~ date + log(cap) + NE + CT + log(N) + PT,
               data = nuclear, family = student(4) )
summary(nuc.t4)
plot(nuc.t4)

# Conditional analysis for partial turnkey guarantee

nuc.t4.pt <- cond( nuc.t4, offset = PT )
summary(nuc.t4.pt)
plot(nuc.t4.pt)

# For conditional analysis for other covariates, replace PT by
# log(N), ...
```

Table 5.3 *Nuclear power station data analysis. Two-sided 95% confidence intervals for coefficients of* `log(N)` *and* `PT`, *under assumed normal and t_4 response distributions. $\widehat{\psi}$ and $\widehat{\psi}_m$ represent respectively the maximum likelihood and maximum marginal likelihood estimates.*

Variable	Error distribution	Type	
`log(N)`	Normal	Exact	$(-0.173, -0.002)$
	Student t_4	$\widehat{\psi}$, normal approx.	$(-0.144, 0.024)$
		$\widehat{\psi}_m$, normal approx.	$(-0.167, 0.022)$
		r	$(-0.142, 0.029)$
		r^*	$(-0.161, 0.026)$
`PT`	Normal	Exact	$(-0.460, 0.008)$
	Student t_4	$\widehat{\psi}$, normal approx.	$(-0.479, -0.084)$
		$\widehat{\psi}_m$, normal approx.	$(-0.480, -0.051)$
		r	$(-0.480, -0.076)$
		r^*	$(-0.483, -0.038)$

order results are based on the maximum likelihood estimates and diagonal elements of the inverse observed information matrix, while the higher order results are the corresponding quantities based on the marginal likelihoods for the individual parameters. There is no consistent pattern to the differences among the results from the analyses, which can be appreciable – in particular, using t_4 errors sharply reduces the significance of `log(N)` and increases that of `PT`. Table 5.3 shows confidence intervals for the coefficients for these two last variables. For both variables, use of higher order approximations lengthens the intervals when the t_4 response distribution is used, and in both cases the intervals fall into two groups of broadly similar intervals, based on first order and on higher order approximations. Use of t_4 responses and higher order approximation yields intervals which are centred differently than is the case for normal responses: for `log(N)` the higher order intervals are somewhat longer than for normal responses, while those for `PT` are of similar length.

It is natural to wonder to what extent the decision to include both `PT` and `log(N)` in the model depends on the assumed response distribution. Table 5.4 shows how the two-sided significance levels for inclusion of these variables depends on the degrees of freedom. The significance levels are calculated as $2\min(p_-, p_+)$, where p_- and p_+ are the one-sided tail probabilities

$$p_- = \Pr(R \leq r; \psi = 0), \quad p_+ = \Pr(R \geq r; \psi = 0)$$

for the first order tests, and

$$p_- = \Pr(R^* \leq r^*; \psi = 0), \quad p_+ = \Pr(R^* \geq r^*; \psi = 0)$$

for the higher order tests. When $\nu = \infty$ these correspond to the use of normal responses with p_- and p_+ computed by applying a normal approximation to the usual t statistic,

Table 5.4 *Nuclear power station data analysis. Significance levels for inclusion of covariates* log(N) *and* PT *as functions of degrees of freedom ν for Student response distribution, in the model with 6 covariates summarized in Table 5.2.*

	log(N)		PT	
ν	First order	Higher order	First order	Higher order
4	0.162	0.151	0.005	0.024
6	0.110	0.116	0.007	0.032
8	0.081	0.098	0.009	0.036
10	0.064	0.086	0.011	0.038
20	0.036	0.064	0.016	0.045
40	0.025	0.053	0.020	0.050
100	0.020	0.047	0.022	0.053
∞	0.035	0.045	0.046	0.057

and using the correct t distribution, respectively. The estimated correlation between the estimated coefficients of PT and log(N) using the normal model of Table 5.2 is -0.6, which may account for the relation between the significance levels as ν varies. If the degrees of freedom ν is small then log(N) seems to be unnecessary, but this conclusion does not depend on the use of higher order procedures.

Figure 5.1 compares the profile and marginal log likelihoods for $\log \sigma$ using Student t errors with $\nu = 4, 20$ degrees of freedom. The downward bias of the ordinary maximum likelihood estimator accounts for the leftward shift of the profile log likelihood relative

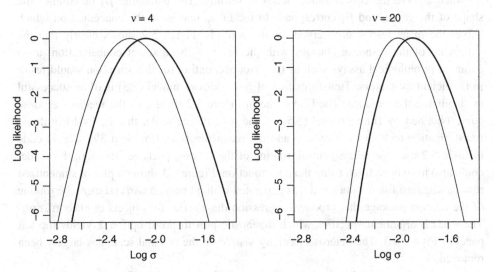

Figure 5.1 Inference for log scale parameter for nuclear power data under the assumption of t_ν response distribution. Profile log likelihoods (solid) and marginal log likelihoods (heavy) for $\log \sigma$, for $\nu = 4$ (left) and $\nu = 20$ (right).

to the marginal log likelihood. As mentioned above, these marginal log likelihoods are also the log posterior densities for $\log \sigma$ obtained when Bayesian inference is performed using the improper prior density proportional to $1/\sigma$. With normal errors σ represents the standard deviation of a single response; with Student errors σ is the scale parameter for the model, but the standard deviation of a single response is $\sigma \sqrt{\nu/(\nu-2)}$; this difference in interpretation should be kept in mind when comparing values of σ across different non-normal distributions.

5.3 *Daphnia magna* data

Chèvre *et al.* (2002) present a study to assess the impact of the herbicide dinoseb on the survival of *Daphnia magna* Straus, 1820, a micro-crustacean widely used as test organism in aquatic ecotoxicological assays. The data, given in Table 5.5, comprise 136 observations and are plotted in Figure 5.2. To enhance readability only concentrations up to 1 mg/l have been plotted; for higher concentrations data points and plotted curves would agree to within drawing accuracy. The data are stored in the R data frame `daphnia` of the `nlreg` package; `time` represents the survival time and `conc` the concentration of dinoseb.

We use the nonlinear regression model (5.3), and model the concentration-response relationship by the four-parameter logistic function

$$\mu(x; \beta) = \beta_1 + \frac{\beta_2 - \beta_1}{1 + (x/\beta_4)^{2\beta_3}}, \qquad x \geq 0, \ \beta_1, \ldots, \beta_4 \geq 0, \qquad (5.5)$$

which results in a sigmoidal curve decreasing from an initial value β_2 to a limiting value β_1 when x, here the concentration, tends to infinity. The parameter β_3 determines the shape of the curve, and β_4 corresponds to the EC_{50}, that is, to the concentration which halves the mean response observed for the controls. Figure 5.2 shows clearly that the variability of the response changes with the level of the toxicant concentration, as is common in biological assays such as this. Not accounting for this variation would result in inefficient estimators. Transformation of both sides of model (5.3) is often successful in stabilizing the variance. The heavy line in Figure 5.2 represents the median response curve obtained by fitting model (5.5) on the logarithmic scale, that is, by logarithmic transformation of both the response and the mean functions (Problem 39). The R code in Code 5.2 uses the `nlreg` fitting routine of the `nlreg` package, although the model could also have been fitted using the `nls` function. Figure 5.3 shows a plot of studentized residuals against fitted values and was generated with `nlreg.diag.plots`, the routine of the `nlreg` package that produces regression diagnostics for objects of class `nlreg`. The `which` argument specifies which diagnostic plot to print; option 2 yields the left panel of Figure 5.3. The heteroscedasticity visible on the original scale has largely been removed.

Reparametrization of the mean function of a nonlinear model is often used for inference on quantities that are expressible as functions of the original parameters. To illustrate

Table 5.5 *Data on the impact of the herbicide dinoseb on the survival of the microcrustacean* Daphnia magna Straus, 1820 *(Chèvre et al., 2002). The tested concentrations range from 0.0 mg/l to 11.3 mg/l. The design is highly unbalanced, there being only 14 concentrations with replicated observations.*

dinoseb concentration (mg/l)											
0	0.006	0.009	0.013	0.02	0.04	0.06	0.09	0.11	0.14	0.16	0.18
survival time (days)											
69	78	86	110	16	9	60	65	12	86	102	114
13				28							
94											
58											
112											
102											
14											

dinoseb concentration (mg/l)											
0.21	0.22	0.23	0.26	0.28	0.32	0.37	0.41	0.45	0.46	0.56	0.66
survival time (days)											
14	3	9	8	8	20	5	3	6	7	0.813	3
	13			10						0.813	
	15			10						1.04	
	20			20						1.04	
	20			48						3	
	35			50						8	
	47			52						13	
	50			52						13	
	52			52						13	
	66			57						13	
	75			65						20	

dinoseb concentration (mg/l)											
0.74	0.92	1.1	1.4	1.7	1.8	1.9	2.3	2.8	3.4	5.7	11.3
survival time (days)											
1.04	0.135	0.385	0.292	0.229	0.0729	0.125	0.0729	0.0729	0.0729	0.0417	0.0208
		0.625	0.115	0.271	0.158		0.108	0.0833	0.0833	0.0521	0.0313
		0.958		0.292			0.146	0.104	0.0833	0.0521	0.0313
		0.958		0.292			0.167	0.125	0.0938	0.0521	0.0313
		0.958		0.396			0.167	0.125	0.0938	0.0521	0.0313
		0.813					0.177	0.125	0.0938	0.0573	0.0313
		0.813					0.188	0.125	0.0938	0.0625	0.0313
		0.813						0.167	0.0938	0.0729	0.0313
		0.813						0.167	0.115	0.0729	0.0313
		0.813						0.188	0.115	0.0729	0.0313
		0.813								0.0833	
		0.813									
		0.813									
		0.813									

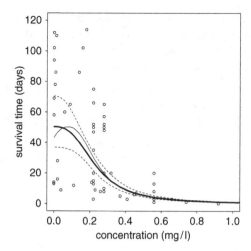

Figure 5.2 *Daphnia magna* data, fitted curves (——, standard logistic; ——, hormesis) and 95% confidence bands for the standard logistic curve, obtained from the r^* pivot.

Code 5.2 R code for analysis of *Daphnia magna* data.

```
> library(nlreg)
> data(daphnia)

# Fit transform both sides (TBS) model with constant variance and
# produce diagnostic plot

> daphnia.nl <-
+     nlreg( log(time) ~ log(b1+(b2-b1)/(1+(conc/b4)^(2*b3))),
+           data = daphnia,
+           start = list(b1 = 0.05, b2 = 50, b3 = 1.5, b4 = 0.2) )
> summary( daphnia.nl )

Regression coefficients:
    Estimate  Std. Error  z value  Pr(>|z|)
b1    0.042      0.008      5.09    3.5e-07
b2   50.412      8.195      6.15    7.7e-10
b3    1.369      0.083     16.41    < 2e-16
b4    0.232      0.028      8.27    < 2e-16

Variance parameters:
       Estimate  Std. Error
logs    -0.661      0.121

No interest parameter

Total number of observations: 136
Total number of parameters: 5
-2*Log Likelihood 296

> nlreg.diag.plots( daphnia.nl, which = 2 )
```

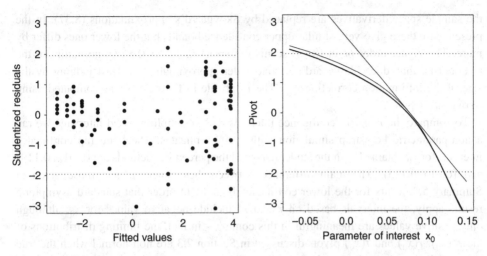

Figure 5.3 *Daphnia magna* data: two examples of graphical output obtained with the routine `nlreg`. `diag.plots` routine and the `profile` method of the `nlreg` package. Left: studentized residuals against fitted values. Right: profile plots for x_0: ——, likelihood root (bent) and Wald pivot (straight); ——, modified likelihood root. The horizontal lines are at ± 1.96.

this, we outline how to calculate the 95% pointwise confidence bands for the median response $\mu(x; \beta)$ that are shown in Figure 5.2. They were obtained from the modified likelihood root (8.32); in this case the confidence bands based upon its first order analogue r agree to within drawing accuracy. To calculate them we reparametrized $\mu(x; \beta)$ as $(\psi, \beta_1, \beta_3, \beta_4)$, where $\psi = \mu(x_+; \beta)$ is the median response for the concentration x_+, and refitted the model for several values of x_+. The confidence bands were obtained by cubic spline interpolation of the upper and the lower confidence bounds for different values of ψ.

One issue of substantive interest here is whether low concentrations of the toxicant have a beneficial effect. This so-called hormesis effect can be included in the model by extending the previous four-parameter logistic function to

$$\mu(x; \beta) = \beta_1 + \frac{\beta_2 - \beta_1}{1 + \{(x - x_0)/\beta_4\}^{2\beta_3}}, \tag{5.6}$$

where x_0 represents the concentration with maximal stimulation. The corresponding fitted curve is plotted in Figure 5.2. With this formulation the interpretation of the parameters β_1, β_2 and β_3 remains the same as above, but the EC_{50} now equals $\beta_4 + x_0$. The right panel of Figure 5.3 gives the profile plot for the parameter x_0 obtained with the `profile` method available for objects of class `nlreg`, which enhances classical first order profile plots; see the bibliographic notes. The corresponding confidence intervals are given in Code 5.3. It appears that confidence intervals based on the Wald statistic are likely to be biased and too short. Two versions of the r^* pivot are available, corresponding to the formulations (8.33) and (8.32) of the q correction term, where in the latter case the sample space derivatives are replaced by Skovgaard's approximations (8.37). In the present case these give very similar upper confidence bounds but the lower ones differ by more. The higher order pivot (8.33) seems to correct more for the non-normality of the test statistic than does Skovgaard's second order approximation, at least judging by the size of the information correction r_{INF}. The r^* profile in Figure 5.3 was computed using Skovgaard's pivot.

To compare the r and r^* confidence intervals with resampling-based intervals, we ran a non-parametric bootstrap simulation with 1,999 replicates. The bootstrap confidence intervals for x_0 obtained with the studentized and the percentile methods are $(0.012, 0.119)$ and $(0.013, 0.123)$. The main difference between these intervals and those given in Summary 5.3 occurs for the lower confidence bounds. In order that standard asymptotic results apply, the intervals based on r and r^* include negative values for x_0, although only positive values are meaningful in this context – in fact, the limiting distributions of the $r^*(x_0)$, $r(x_0)$ and $t(x_0)$ pivots discussed in Section 2.3 are non-normal when the true value of the parameter lies on the boundary of the parameter space; see the bibliographic notes.

A further question of interest is whether there is a threshold effect: is there a concentration below which no effect on the survival time is seen? Details of this analysis are given in Problem 40.

A by-product of using (5.3) on the logarithmic scale is the calculation of a widely used measure of the action of the toxicant. On the original scale the model fitted to the *Daphnia magna* data has a multiplicative error structure with errors that are log-normally distributed and where the median response at the concentration x_i equals $\mu(x_i; \beta)$ (Problem 39). Thus for a given concentration x we can estimate the time $y_{0.5}$ after which 50% of the population will have died,

$$y_{0.5} = \{y \geq 0 \mid \Pr(Y \leq y; x, \widehat{\theta}) = 0.5\} = \mu(x; \widehat{\beta}),$$

where $\widehat{\theta} = (\widehat{\beta}, \widehat{\sigma})$. Conversely, we can calculate the concentration that will cause a mortality of 50% in the population by a specified time y_0:

$$\text{LC}_{50} = \{x \geq 0 \mid \Pr(Y \leq y_0; x, \widehat{\theta}) = 0.5\} = \{x \geq 0 \mid \mu(x; \widehat{\beta}) = y_0\};$$

this is known in toxicology as the 50% lethal concentration. Common values for the observation time are 24 hours and 48 hours – the usual periods considered in acute toxicity tests – and 21 days, the usual duration of chronic toxicity tests. The left-hand panel in Figure 5.4 shows the values of LC_{50} for different time points, together with the 95% pointwise confidence bands. The latter were calculated following the procedure used for calculating the confidence bands for the median response, except that a non-parametric bootstrap was used to replace the likelihood pivot. The specific values for 24 hours, 48 hours and 21 days are also reported. We may also consider other concentrations of

Code 5.3 R code for analysis of *Daphnia magna* data (cont.).

```
# TBS model fit, profile plot and 95% confidence intervals for hormesis
# parameter

> daphnia.h.nl <-
+     nlreg( log(time) ~ log(b1+(b2-b1)/(1+(((conc-x0)/b4)^2)^b3)),
+          start = list( b1 = 0.05, b2 = 50, b3 = 3, b4 = 0.1, x0 = 0),
+          data = daphnia, hoa = TRUE )
> coef( daphnia.h.nl )
    b1    b2    b3    b4      x0
0.0373 50.4 1.190 0.168 0.0785

> daphnia.h.prof <- profile( daphnia.h.nl, offset = x0 )
> plot( daphnia.h.prof )
Higher order method used: Skovgaard's r*

> summary( daphnia.h.prof )

Two-sided confidence intervals for x0
                lower   upper
r* - Fr (0.95) -0.0667  0.119
r* - Sk (0.95) -0.0519  0.121
r (0.95)       -0.0386  0.121
Wald (0.95)     0.0179  0.139

      Estimate  Std. Error
x0    0.079       0.0309

22 points calculated exactly
50 points used in spline interpolation

INF (Sk): 0.185
INF (Fr): 0.401
NP (Sk): 0.273
NP (Fr): 0.275
```

interest, such as those causing the death of 5% or 10% of the population in a given time period:

$$\mathrm{LC}_p = \{x \geq 0 \mid \mathrm{Pr}(Y \leq y_0; x, \widehat{\theta}) = p\} = \{x \geq 0 \mid \mu(x; \widehat{\beta})e^{\widehat{\sigma}z_p} = y_0\},$$

where z_p is the p quantile of the standard normal distribution with $0 < p < 1$. If $p = 0.05$ or $p = 0.1$, these yield the LC_5 and LC_{10}, which are shown in the middle and right-hand panels of Figure 5.4, with their 95% pointwise confidence bands.

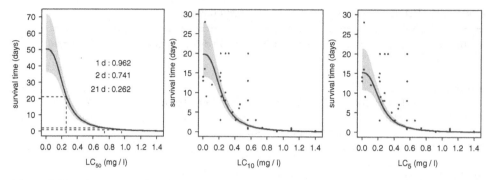

Figure 5.4 *Daphnia magna* data: 50%, 10% and 5% lethal concentrations (mg/l) with 95% pointwise confidence bands (shaded). The left-hand panel shows how the specific values reported in the legend for the time periods 24 hours (1day), 48 hours (2 days) and 21 days are obtained (dashed lines). The dots in the middle and right-hand panels represent the original observations.

5.4 Radioimmunoassay data

Figure 5.5 shows the results of a radioimmunoassay (RIA) taken to estimate the concentrations of a drug in samples of porcine serum (Belanger *et al.*, 1996, Table 1, first two columns). The experiment consists of 16 observations made at eight different drug levels with two replications at each level. The data, shown in Table 5.6, are available in the data frame `ria` of the `nlreg` package; `count` represents the observed percentage of radioactive gamma counts, and `conc` the drug concentration (ng/ml). We model the concentration–response relationship by means of a slightly simplified version of the four-parameter logistic function (5.5),

$$\mu(x; \beta) = \beta_1 + \frac{\beta_2 - \beta_1}{1 + (x/\beta_4)^{\beta_3}}, \quad x \geq 0, \ \beta_1, \dots, \beta_4 \geq 0.$$

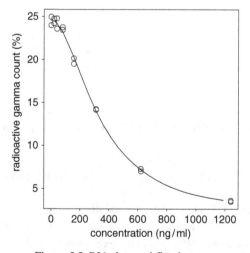

Figure 5.5 RIA data and fitted curve.

Table 5.6 *Results of a radioimmunoassay taken to estimate the concentrations of a drug (ng/ml) in samples of porcine serum (Belanger et al., 1996).*

Concentration							
0.0	19.4	38.8	77.5	155.0	310.0	620.0	1240.0
Radioactive gamma counts (%)							
23.991	24.213	23.596	23.483	19.504	14.242	7.030	3.624
24.973	24.769	24.800	23.770	20.172	14.179	7.299	3.488

Figure 5.5 suggests that the variance of the response might slightly decrease with the response level.

An experimental setting rarely suggests use of a particular variance function, but the dynamic of a radioimmunoassay is well understood and the variance of the associated error distribution may be captured by a power-of-the-mean variance function with $w(x; \beta, \gamma)^2 = \mu(x; \beta)^\gamma$, where γ is a scalar variance parameter. Code 5.4 gives the R code for analyzing the RIA data set. If the variance is constant, we may use either the `nlreg` routine or the native R routine `nls` to fit the model; both yield the same estimates. However, `nls` cannot be used if error heteroscedasticity is modelled using a parametric variance function; in this case the `weights` argument to `nlreg` defines the variance function of the nonlinear model, with the constant parameter σ^2 included by default. The maximum likelihood estimates and standard errors are given in Code 5.4.

The calculation of higher order solutions requires the symbolic differentiation of the mean and variance functions, which is accomplished using the two workhorse functions `Dmean` and `Dvar` – see Chapter 9. This can be time-consuming if the model contains several parameters, and the user can save the first and second derivatives to the fitted model object by setting `hoa = TRUE` in the call to `nlreg`. This makes little difference to the execution time for simple models but appreciably reduces it for more complex variance functions, especially those that depend on the regression coefficients, as is the case for the herbicide case study of Section 6.4.

It is often useful to represent a fitted nonlinear model graphically by using profile pair sketches, which are the bivariate extension of profile plots. The `contour` method for objects of class `nlreg` does this, and also includes higher order solutions. Given two parameters of interest, the method traces an approximation to their joint confidence regions of level $(1 - \alpha)$ obtained from the Wald pivot, the likelihood ratio w, and its higher order version w^* given at (8.61). Code 5.4 shows an application of `contour`, and Figure 5.6 shows its output. The plots are organized as a matrix whose main diagonal contains the profile plots for the parameters, where r^* uses Skovgaard's approximation to q_1; see Section 8.6.3. The bivariate contour plots in the lower triangle are plotted on the original scale, whereas those in the upper triangle are on the r scale; their units are those of the associated likelihood roots, thus removing parameter-dependent components of the curvature – see the bibliographic

Code 5.4 R code for analysis of RIA data.

```
library(nlreg)
data(ria)

# Fit nonlinear model to RIA data

> ria.nl <- nlreg( formula = count ~ b1+(b2-b1)/(1+(conc/b4)^b3),
+                  weights = ~ (b1+(b2-b1)/(1+(conc/b4)^b3))^g,
+                  data = ria, hoa = TRUE,
+                  start = c(b1=1.6, b2=20, b3=2, b4=330, g=2) )
> summary( ria.nl )

Regression coefficients:
     Estimate  Std. Error  z value  Pr(>|z|)
b1      1.802       0.197     9.16     <2e-16
b2     24.575       0.206   119.52     <2e-16
b3      1.899       0.074    25.61     <2e-16
b4    334.956       5.770    58.05     <2e-16

Variance parameters:
       Estimate  Std. Error
g         2.095       0.527
logs     -8.094       1.467

No interest parameter

Total number of observations: 16
Total number of parameters: 6
-2*Log Likelihood 6.527

# Profile plots and profile pair sketches

> ria.prof <- profile( ria.nl )
> par( mar = c(1.5, 1, 2, 1) )
> contour( ria.prof, alpha = 0.05, cl1 = "black", cl2 = "black",
           lwd2 = 2 )
Higher order method used: Skovgaard's r*
```

---

notes. The more elliptical are the contours, the more quadratic is the likelihood, as the pair (β_2, β_3) shows. If the first order statistics behave well, then the curves for w and w^* will be close, but it is striking how much they differ for the RIA data. For most pairs profile traces have been added, representing the constrained maximum likelihood estimates of one parameter as a function of the other. These give useful information on how the estimates affect each other: if the asymptotic correlation is zero, the angle between the traces is close to $\pi/2$, while an angle close to zero, as for the estimates of the variance parameters (γ, σ^2), indicates strong correlation. In this situation the algorithm may fail: for instance, it was impossible to compute all the profile pair sketches for (γ, σ^2), so the contours of

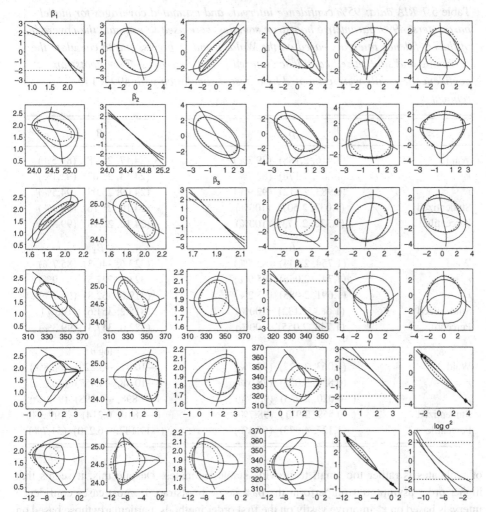

Figure 5.6 RIA data: profile and approximate bivariate contour plots obtained using the `contour` method of the `nlreg` package with $\alpha = 0.05$ (——, w/r; ——, w^*/r^*; - - -, Wald). See explanation starting on page 73.

w^* are missing in the corresponding panels and four bullets indicate where they would intersect the profile traces. The calculation of profile pair sketches is computationally very demanding, as the model has to be refitted many times to obtain the constrained estimates.

Table 5.7 reports the 95% confidence intervals for all parameters, obtained from a summary of the `ria.prof` object. To assess their reliability we ran a parametric bootstrap simulation with 4,999 replicate data sets generated from the fitted model obtained by maximum likelihood estimation, and computed the coverages of the higher order intervals; these are given in Table 5.7. We also calculated the 95% percentile and studentized bootstrap confidence intervals and estimated their coverages using a nested bootstrap simulation with 499×149 replicates; these results should be treated gingerly as the number

Table 5.7 *RIA data: 95% confidence intervals and estimated coverages for model parameters. 4999 parametric bootstrap samples were used to estimate the coverage of the confidence intervals based on the Wald, r and r* pivots, and to calculate the percentile and studentized confidence intervals. The coverages of the bootstrap confidence intervals were estimated by a nested simulation with 499×149 replicates.*

	β_1	β_2
Wald	(1.416, 2.188) (79.3%)	(24.17, 24.98) (89.5%)
r	(1.303, 2.178) (82.7%)	(24.14, 25.02) (92.1%)
r*	(0.983, 2.312) (92.9%)	(24.09, 25.08) (94.5%)
Studentized	(1.037, 2.507) (100%)	(24.10, 25.16) (99.3%)
Percentile	(1.326, 2.173) (100%)	(24.12, 25.98) (99.3%)
	β_3	β_4
Wald	(1.754, 2.044) (83.7%)	(323.6, 346.3) (85.7%)
r	(1.747, 2.056) (85.9%)	(323.0, 347.5) (87.9%)
r*	(1.691, 2.085) (93.2%)	(319.6, 352.8) (93.7%)
Studentized	(1.683, 2.132) (100%)	(318.2, 351.6) (100%)
Percentile	(1.744, 2.055) (100%)	(322.6, 347.3) (99.3%)
	γ	$\log \sigma^2$
Wald	(1.063, 3.127) (72.7%)	(−10.97, −5.220) (66.8%)
r	(0.859, 3.025) (78.3%)	(−10.48, −4.457) (76.0%)
r*	(−0.020, 2.923) (93.3%)	(−9.898, −1.398) (93.6%)
Studentized	(−0.559, 2.971) (98.7%)	(−9.810, 0.214) (99.3%)
Percentile	(1.127, 4.973) (65.1%)	(−17.147, −6.230) (51.7%)

of replicates is rather too small to provide stable estimates. Despite this, it seems that higher order asymptotics outperform the parametric bootstrap. In most cases confidence intervals based on r^* improve vastly on the first order methods, particularly those based on the Wald pivot; the higher order intervals seem to be accurate enough for use in practice. Furthermore, they are much quicker to obtain than are those based on resampling.

Estimation of the variance parameters in nonlinear heteroscedastic regression is vexed. Maximum likelihood estimators are often very biased, as suggested by the profile plots in Figure 5.6. Better estimators can be obtained from the adjusted profile likelihood, for example by using the fitting routine `mpl` included in the `nlreg` package, which yields $\widehat{\gamma}_m = 1.79$ (0.703) and $\log \widehat{\sigma}_m^2 = -6.95$ (2.027); the standard errors here are based upon the profile information matrix. The R code for this computation is given in Code 5.5. The `summary` output of `mpl` objects gives also the constrained maximum likelihood estimates of the regression coefficients, where the variance parameters are estimated from the adjusted profile log likelihood. Figure 5.7 compares the profile and the adjusted profile log likelihoods. The shaded regions in the horizontal plane represent the 95% bivariate confidence regions obtained from the likelihood ratio and the modified likelihood ratio statistic w^*. They agree with the approximate contour plots for $(\gamma, \log \sigma^2)$ shown in

Code 5.5 R code for analysis of RIA data.

```
# Compute maximum adjusted profile likelihood estimates of variance
# parameters

> ria.mpl <- mpl( ria.nl )
> summary( ria.mpl )
Higher order method used: Skovgaard's r*

Variance parameters
         MMPLE     MLE   Std.   Error
g        1.785   2.095          0.703
logs    -6.947  -8.094          2.027

Regression coefficients
       MMPLE      MLE   Std.   Error
b1     1.780    1.802          0.248
b2    24.583   24.575          0.224
b3     1.891    1.899          0.086
b4   335.017  334.956          7.036

Total number of observations: 16
Total number of parameters: 6
-2*Log Lmp 17.54
```

PROFILE LIKELIHOOD

(MLEs and 95% confidence region added)

ADJUSTED PROFILE LIKELIHOOD

(MLEs and 95% confidence region added)

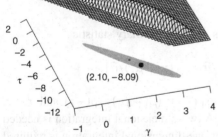

Figure 5.7 RIA data: profile and adjusted profile log likelihoods for the variance parameters γ and $\tau = \log \sigma^2$, maximum likelihood estimates, and corresponding 95% confidence regions. Based on the output of the `mpl` fitting routine of the `nlreg` package.

the corresponding panel of Figure 5.6 (last row, second from the right). The two points show the maximum likelihood and maximum adjusted likelihood estimates, the distance between which gives an idea of the bias of the usual maximum likelihood estimator.

5.5 Leukaemia data

Table 5.8 gives survival times in weeks of 17 patients with leukaemia, along with their white blood cell counts at the time of diagnosis. Cox and Snell (1981, Example U) use these data to illustrate exponential regression with independent responses, their densities being

$$f(t_i; x_i, \beta) = \lambda_i^{-1} \exp(-t_i/\lambda_i), \quad i = 1, \ldots, 17,$$

where

$$\lambda_i = \beta_0 \exp\{\beta_1(x_i - \bar{x})\} \tag{5.7}$$

and x_i is the base 10 logarithm of the white blood cell count. This model fits well, and shows that the mean survival time decreases as the white blood cell count increases.

We use these data to illustrate higher order approximations for a Weibull model, and to show how this is easily extended to Type II censoring, which we artificially create. The Weibull distribution function is

$$F(t_i; \lambda_i, \kappa) = 1 - \exp\left\{-\left(\frac{t_i}{\lambda_i}\right)^\kappa\right\}, \quad t_i > 0, \quad \lambda_i, \kappa > 0,$$

from which one can verify that $y_i = \log t_i$ follows a regression-scale model

$$y_i = \log \beta_0 + \beta_1(x_i - \bar{x}) + \sigma \varepsilon_i,$$

where β_0 and β_1 are the parameters appearing in (5.7), $\sigma = 1/\kappa$, and $f(\varepsilon) = \exp(\varepsilon - e^\varepsilon)$ is the density of a log-Weibull variable, sometimes called the extreme-value density.

The log likelihood function for β_0, β_1, and $\tau = \log \sigma$ may be written as

$$\ell(\beta, \tau; y) = -\left\{n\tau + \sum_{i=1}^n [\exp\{(y_i - \mu_i)e^{-\tau}\} - (y_i - \mu_i)e^{-\tau}]\right\},$$

where $\mu_i = \log \lambda_i = \log \beta_0 + \beta_1(x_i - \bar{x})$. There is an exact ancillary statistic

$$a = (a_1, \ldots, a_n) = \{(y_1 - \hat{\mu}_1)/\hat{\sigma}, \ldots, (y_n - \hat{\mu}_n)/\hat{\sigma}\},$$

and the marginal densities of $(\log \widehat{\beta}_0 - \log \beta_0) \exp(-\hat{\tau})$, $(\widehat{\beta}_1 - \beta_1) \exp(-\hat{\tau})$ and $\hat{\tau} - \tau$, conditional on a, can be obtained by integration. A two-dimensional integration is needed to eliminate the nuisance parameter, and a further one-dimensional integration is required to norm the resulting density.

The components of r^* required for inference about β_1 can be obtained from (8.49) as

$$r = r_p(\mu) = \text{sign}(q) \left[2\left\{\ell_p(\widehat{\beta}_1) - \ell_p(\beta_1)\right\}\right]^{1/2}, \tag{5.8}$$

$$q = q(\mu) = \ell_p'(\beta_1)|j_{\lambda\lambda}(\beta_1, \widehat{\lambda}_{\beta_1})|^{1/2}|j(\widehat{\beta}, \widehat{\lambda})|^{-1/2}, \tag{5.9}$$

Table 5.8 *Survival time t in weeks of leukaemia patients, and associated log$_{10}$ white blood cell count x (Feigl and Zelen, 1965), after Cox and Snell (1981, Example U).*

x	t	x	t	x	t
3.36	65	4.00	121	4.54	22
2.88	156	4.23	4	5.00	1
3.63	100	3.73	39	5.00	1
3.41	134	3.85	143	4.72	5
3.78	16	3.97	56	5.00	65
4.02	108	4.51	26		

Table 5.9 *Limits of first order and third order confidence intervals for β_1 based on the data in Table 5.8.*

	1%	5%	10%	90%	95%	99%
			Weibull model			
Wald	−1.941	−1.790	−1.660	−0.326	−0.196	−0.046
r	−2.121	−1.906	−1.734	−0.330	−0.191	−0.021
r*	−2.273	−2.031	−1.839	−0.293	−0.138	0.053
			Exponential model			
Wald	−1.912	−1.766	−1.640	−0.344	−0.217	−0.071
r	−1.909	−1.762	−1.636	−0.334	−0.203	−0.050
r*	−1.909	−1.762	−1.635	−0.326	−0.194	−0.040

where $\lambda = (\log \beta_0, \tau)$ is the two-dimensional nuisance parameter. Similar expressions are available for inference about β_0 and τ.

Table 5.9 shows first order and higher order confidence intervals for the slope parameter β_1, under both the exponential and the Weibull models. There are slight differences, but all methods lead to the conclusions that the slope of the regression is approximately -1 and that the relationship between failure time and white blood cell count is statistically significant. Code 5.6 gives the code that was used to generate this table.

Type II censoring is the name given to observing failure times of n items until a fixed number, r, of these items have failed. It arises in some reliability applications, but random censoring is more common in studies of human populations. The conditional analysis of regression-scale models can be directly extended to Type II censoring. For the log-Weibull model, the log likelihood function becomes

$$\ell(\beta, \tau; y) = - \left[r\tau - \sum_{i=1}^{r} (y_{(i)} - \mu_{(i)}) e^{-\tau} + \sum_{i=1}^{n} \exp \left\{ (y_{(i)} - \mu_{(i)}) e^{-\tau} \right\} \right],$$

Code 5.6 R code used for analysis of leukemia data.

```
# function to compute minus log likelihood for example U ("exU") of
# Cox and Snell (1981), including some censoring if appropriate

> llweibreg <- function(logb0 = 4, b1 = -1, tau = 0) {
+    y <- log( exU$time[exU$status==1] )
+    mu1 <- logb0 + b1 * exU$x[exU$status==1]
+    t1 <- sum( tau + exp((y-mu1) * exp(-tau)) - (y-mu1) * exp(-tau) )
+    c <- log( exU$time[exU$status==0] )
+    mu2 <- logb0 + b1 * exU$x[exU$status==0]
+    t2 <- sum( exp((c-mu2) * exp(-tau)) )
+    t1 + t2 }

# maximum likelihood fit to original data (no censoring)

> library(stats4)
> ( full.fit <- mle(llweibreg) )

Coefficients:
        Estimate  Std.  Error
logb0    3.9568         0.2569
b1      -0.9934         0.4066
tau      0.0036         0.2011

# This is the loop for profiling

> logj2 <- b0.b1 <- profloglik.b1 <- vector( "numeric",
+                                    length = length(b1.v))
> for( i in seq(along = b1.v) ) {
+    fit <- mle( llweibreg, fixed = list(b1=b1.v[i]) )
+    logj2[i] <- - log( det(vcov(fit)) )
+    b0.b1[i] <- coef(fit)[1]
+    profloglik.b1[i] <- as.numeric( logLik(fit) ) }

# Higher order calculations

> profloglik.b1 <- profloglik.b1 - max(profloglik.b1)
> lp1 <- predict( smooth.spline(b1.v, profloglik.b1), b1.v, 1 )
> rho <- 0.5 * ( log( det(vcov(full.fit)) ) + logj2 )
> q <- lp1$y * exp(rho)
> r.p <- sign(q) * sqrt( -2 * profloglik.b1 )
> rstar <- r.p + (1/r.p) * log(q/r.p)

# Confidence intervals by spline smooths avoiding middle for r*

> rstarshort <- c( rstar[1:94], rstar[106:200] )
> b1.vshort <- c( b1.v[1:94], b1.v[106:200] )
> predict( smooth.spline(rstarshort, b1.vshort),
          c(2.33, 1.96, 1.64, -1.64, -1.96, -2.33), 0 )$y
[1] -2.27346665 -2.03122085 -1.83886173 -0.29335091 -0.13854009
[6]  0.05301845
```

Table 5.10 *Effect of artificial Type II censoring on inference for β_1 based on the data of Table 5.8.*

	$\widehat{\beta}_0$	$\widehat{\beta}_1$	95% CI for β_1 (r^*)	$\widehat{\sigma}$
No censoring	52.29	−0.99	(−2.031, −0.138)	1.004
$r = 15$	57.47	−1.27	(−2.636, −0.305)	1.107
$r = 13$	64.30	−1.47	(−3.275, −0.358)	1.212
$r = 11$	63.67	−1.46	(−3.818, −0.301)	1.231

where we have written the r observed failure times in order, $y_{(1)} \leq \cdots \leq y_{(r)}$, and $\mu_{(i)}$ is the mean associated with $y_{(i)}$; the final term in the log likelihood function includes the further $(n - r)$ observations censored at $y_{(r)}$. As in the uncensored case there is an exact ancillary statistic

$$a = (a_1, \ldots, a_r) = \{(y_{(1)} - \widehat{\mu}_{(i)})e^{-\widehat{\tau}}, \ldots, (y_{(r)} - \widehat{\mu}_{(r)})e^{-\widehat{\tau}}\}$$

and expressions (5.8) and (5.9) can be used for higher order analysis.

We introduce artificial Type II censoring for the data in Table 5.8 by supposing first that observation continued until there were 15 failures – thus the largest observed value is 134 weeks – then assuming there were 13 observed failures, giving $y_{(r)} = 108$, and finally 11 observed failures, with $y_{(r)} = 65$. The `marg` package does not at present include an option for censoring, but the code in Code 5.6 can be used with minor changes. Point estimates and 95% confidence intervals for β_1 are given in Table 5.10.

5.6 PET film data

Table 5.11 shows failure time data from an accelerated life test on PET film in gas insulated transformers; the film is used as insulation in electrical products. The four different voltages x have a large effect on the failure time. Three observations at $x = 5$ are censored: according to the data source they were censored at a pre-determined time but their values suggest otherwise and we shall suppose below that Type II censoring was applied. A bootstrap analysis of these data is presented in Davison and Hinkley (1997, pp. 346–350), using the Weibull failure time distribution

$$F(y; \gamma, \kappa) = 1 - \exp\{-(y/\gamma)^\kappa\}, \quad y > 0, \ \gamma, \kappa > 0.$$

We shall suppose that the scale and shape parameters depend on x according to the prescription

$$\gamma = \exp\{\beta_0 + \beta_1 \log(x - 5 + e^{\beta_4})\}, \tag{5.10}$$

$$\kappa = \exp(\beta_2 + \beta_3 \log x); \tag{5.11}$$

this allows the *threshold*, the x-value below which failure cannot occur, to be unknown. The threshold is a non-regular parameter, so the usual asymptotic theory does not apply

Table 5.11 *Failure times from an accelerated life test on PET film in gas insulated transformers operated at different voltages (Hirose, 1993). Three observations marked by* + *were right-censored.*

Voltage (kV)	Failure times (hours)					
5	7131	8482	8559	8762	9026	9034
	9104	9104.25+	9104.25+	9104.25+		
7	50.25	87.75	87.76	87.77	92.90	92.91
	95.96	108.30	108.30	117.90	123.90	124.30
	129.70	135.60	135.60			
10	15.17	19.87	20.18	21.50	21.88	22.23
	23.02	23.90	28.17	29.70		
15	2.40	2.42	3.17	3.75	4.67	4.95
	6.23	6.68	7.30			

to its estimation, although we may still be able to treat it as a nuisance parameter and use higher order procedures for inference on the other parameters.

The interest parameter investigated by Davison and Hinkley (1997) is the mean failure time at voltage $x_0 = 4.9$ kV, which may be written as $\psi = 10^{-3} \gamma_0 \Gamma(1 + 1/\kappa_0)$, where γ_0 and κ_0 are (5.10) and (5.11) evaluated with $x = x_0$ and $\Gamma(\cdot)$ represents the gamma function. In our general notation we may write

$$\beta_0 = \log \psi + 3 \log 10 - \lambda_1 \log(x_0 - 5 + e^{\lambda_4}) - \log \Gamma(1 + e^{-\lambda_2 - \lambda_3 \log x_0}),$$

and $\beta_1 = \lambda_1$, $\beta_2 = \lambda_2$, $\beta_3 = \lambda_3$, and $\beta_4 = \lambda_4$. Higher order inference for ψ may be based on expressions (2.11) and (2.12). The quantities V_i are obtained by applying (2.14) to the pivotal quantities $z_i = (y_i/\lambda_i)^{\kappa_i}$, which have standard exponential distributions. This yields

$$V_i = y_i \times \left(1, \ \log(x_i - 5 + e^{\widehat{\lambda}_4}), \ -\log\left(\frac{y_i}{\widehat{\gamma}_i}\right), \ -\log\left(\frac{y_i}{\widehat{\gamma}_i}\right) \log x_i, \ \frac{\widehat{\lambda}_1 e^{\widehat{\lambda}_4}}{x_i - 5 + e^{\widehat{\lambda}_4}}\right),$$

where $\widehat{\lambda}_1$ and $\widehat{\lambda}_4$ are overall maximum likelihood estimates and $\widehat{\gamma}_i$ is the fitted value of γ for the ith observation. We also require

$$\frac{\partial \ell(\theta)}{\partial y_i} = y_i^{-1} \left\{ \delta_i(\kappa_i - 1) - \kappa_i (y_i/\gamma_i)^{\kappa_i} \right\}, \quad i = 1, \ldots, n,$$

where $\delta_i = 0$ or 1 according as to whether y_i is censored or not, and γ_i, κ_i are the scale and shape parameters, evaluated at $\theta = (\psi, \lambda_1, \ldots, \lambda_4)$ and $x = x_i$.

In this case the evaluation of (2.11) close to the overall maximum likelihood estimate is particularly delicate, though the limits of 95% confidence intervals seem to be quite well determined. Figure 5.8 shows pivot profiles for ψ based on the Wald pivot t, the likelihood root r and the modified likelihood root r^* as well as the limits

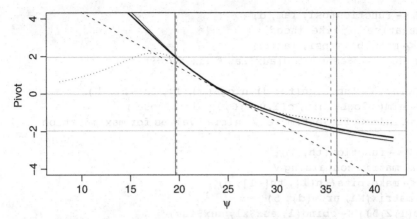

Figure 5.8 PET film data: pivot profiles using the Wald pivot t (dashes), the likelihood root r (solid) and the modified likelihood root r^* (bold). Also shown are the quantity q (dotted), the limits of the 95% confidence intervals based on the likelihood ratio statistic (vertical lines, grey dashes) and the likelihood ratio statistic calibrated using parametric bootstrap sampling (vertical lines, grey solid). The grey horizontal lines are at $0, \pm 1.96$.

of the 95% bootstrap confidence interval. The corresponding 95% confidence limits for ψ are $(18.15, 31.50)$ thousand hours, $(19.75, 35.53)$ thousand hours, $(19.67, 37.34)$ thousand hours, and the bootstrap interval is $(19.62, 36.12)$ thousand hours; this last is obtained by using a parametric bootstrap simulation based on the fitted Weibull model to calibrate the likelihood ratio statistic for ψ. As usual the interval based on the Wald pivot is inappropriately symmetric around the maximum likelihood estimate and the other intervals account for the asymmetry of the log likelihood around its maximum, though to different extents. The code for the higher order analysis is given in Code 5.7.

Code 5.7 R code for analysis of PET film data.

```
make.bits <- function(psi, lam, d) {
#   auxiliary function to save space below
 xl <- log( d$volt - 5 + exp(lam[4]) )
 x0 <- 4.9
 xl0 <- log( x0 - 5 + exp(lam[4]) )
 kap <- exp( lam[2] + lam[3]*log(d$volt) )
 kap0 <- exp( lam[2] + lam[3]*log(x0) )
 beta0 <- log(psi) + 3*log(10) - lam[1]*xl0 - lgamma(1+1/kap0)
 gam <- exp( beta0 + lam[1]*xl )
 y <- d$time
 z <- (y/gam)^kap
 list( z = z, gam = gam, kap = kap, xl = xl,
       inv = lam[1] * exp(lam[4]) / (d$volt - 5 + exp(lam[4])) ) }
```

```
nlogL <- function(psi, lam, d) {
#   negative log likelihood
 aux <- make.bits(psi, lam, d)
 sum( aux$z - d$cens * log(aux$kap * aux$z / d$time) ) }

nlogL.init <- function(th, d) nlogL( th[1], th[-1], d )
fit <- nlm( nlogL.init, c(1,0,0,0,0), d = hirose )
hirose.init <- fit$estimate # initial values for maximisation

make.V <- function(th, d) {
#  make matrix containing V
 aux <- make.bits(th[1], th[-1], d)
 V <- matrix(NA, nrow(d), 5)
 V[,c(1,2,5)] <- cbind(1, aux$xl, aux$inv)
 V[,3] <- - log(d$time / aux$gam )
 V[,4] <- - log(d$time / aux$gam) * log(d$volt)
 V * d$time }

phi <- function(th, V, d) {
#  use V to compute local parametrization
  aux <- make.bits(th[1], th[-1], d)
  L <- (d$cens * (aux$kap-1) - aux$z * aux$kap) / d$time
  apply(L*V, 2, sum) }

library(boot)
data(hirose)

hirose.fr <- fraser.reid( NULL, nlogL = nlogL, phi = phi, make.V = make.V,
                    th.init = jitter(hirose.init), d = hirose )
plot.fr(hirose.fr)
lik.ci(hirose.fr)
```

Bibliographic notes

Exact conditional inference for linear regression with non-normal errors builds on the analysis of the location-scale model first outlined in Fisher (1934) and is discussed in Fraser (1979, Chapter 6). The higher order approximation used in the `marg` package is derived by DiCiccio *et al.* (1990). The use of Student t errors as a robust alternative to least squares is discussed by Lange *et al.* (1989), but they did not investigate the effect of conditioning on the exact ancillary. Higher order inference in linear and nonlinear regression using the general approach of Section 8.5.3 and expression (8.33) is discussed by Fraser *et al.* (1999b). Davison (2003, Section 4.6) discusses likelihood asymptotics in non-regular models.

The use of AIC and AIC_c for model selection is discussed by Hurvich and Tsai (1989, 1991); AIC_c may be motivated by small-sample considerations akin to those underlying the development of higher order asymptotics. McQuarrie and Tsai (1998) give a comprehensive discussion of these and related criteria.

The nonlinear regression examples are taken from Brazzale (2000, Chapter 5), and the higher order procedure used is based on the approximate ancillarity method of Skovgaard (1996); see also the bibliographic notes for Chapter 8. Bellio *et al.* (2000) describe the use of Skovgaard's approximation in nonlinear regression, while Bellio (2000, Chapter 2) emphasises regression diagnostics in the same setting.

Nonlinear least squares is surveyed in Bates and Watts (1988). Section 6.1 of their book discusses profile plots and traces, and the `nlreg` package implements these in the `profile` method. Profile pair sketches, illustrated in Figure 5.6, extend the algorithm of Bates and Watts (1988, Appendix A.6). The studentized and percentile bootstrap methods applied in Section 5.3 are described in Davison and Hinkley (1997, Chapter 5).

Several methods of fitting model (5.4) are summarized in Chapter 2 of Seber and Wild (1989). Perhaps the most commonly used of these is generalized nonlinear least squares, which is surveyed by Carroll and Ruppert (1988, Chapter 2). First order inference and linearization techniques are described in Seber and Wild (1989, Chapter 5). Bellio and Brazzale (1999) show that nonlinearity of the mean function and variance heterogeneity can lead to serious inaccuracies if the sample size is small or moderate.

Lawless (2003, Chapter 5) describes exact conditional inference for censored data, following the general approach outlined in Fraser (1979, Chapters 2–4). Higher order approximations for Type II censored data are considered in DiCiccio *et al.* (1990) and Wong and Wu (2000).

6

Some case studies

6.1 Introduction

In Chapters 4 and 5, we presented a selection of case studies with the goals of emphasizing the application and of illustrating higher order approximations as an adjunct to inference. The selection of our case studies was informed by these twin goals – we used relatively small data sets, and our discussion was sometimes rather remote from the original application.

In this chapter we present more detailed analyses of data collected to address particular scientific problems, with emphasis on the modelling and the conclusions. While we use higher order methods as an adjunct to the analysis, the main focus is on the data analysis rather than the inference methods. These case studies are a subset of examples that have crossed our desks in collaborative work. In Chapter 7 we take the opposite approach, and illustrate a selection of inference problems that are amenable to higher order approximation.

Sections 6.2 and 6.3 present slightly non-standard analysis of binary data; in the first case a natural model leads to the complementary log–log link, and in the second we consider conditional assessment of the binary model, eliminating the parameters in the binary regression by conditioning on the sufficient statistic. In Section 6.4 we present detailed analysis of a published set of herbicide data, with particular emphasis on the `nlreg` package of the `hoa` bundle. This package provides an extensive set of diagnostics and plots that are a useful adjunct to first order analysis, as well as providing an implementation of higher order approximations.

6.2 Wasp data

Table 6.1 summarizes data from an experiment conducted by Britta Tschanz and Sven Bacher of the Zoological Institute at the University of Bern on the behaviour of the paper wasp *Polistes dominulus* in the presence of their prey, larvae of the shield beetle *Cassida rubiginosa*. The purpose of the experiment was to shed light on the behaviour of a predator–prey system in relatively natural conditions, with neither prey nor predators confined during predation; it seems to be one of the first experiments to do so.

Table 6.1 *Summary of wasp data, giving the number of cages open on each of 32 days, with r out of m larvae eaten. The estimated number of wasps active each day was n_w.*

Month	Day	Cages	n_w	m	r	Month	Day	Cages	n_w	m	r
6	26	3	67	20	16	7	27	1	58	30	27
6	30	1	57	10	10	7	28	2	62	30	28
7	6	0	42	5	1	7	29	0	51	15	15
7	10	0	43	20	11	7	30	2	59	20	19
7	11	1	53	35	24	7	31	1	55	25	24
7	14	2	56	35	33	8	1	3	69	30	30
7	15	3	66	5	5	8	2	0	52	40	37
7	16	3	69	40	39	8	3	3	62	10	10
7	17	0	50	25	22	8	4	0	49	30	28
7	18	2	56	10	9	8	5	1	54	20	17
7	19	1	55	40	37	8	7	2	59	25	24
7	20	3	59	15	12	8	8	3	68	25	24
7	21	2	55	40	35	8	9	2	55	5	5
7	22	0	49	35	28	8	10	1	50	5	5
7	23	1	54	15	12	8	11	2	61	15	15
7	26	3	68	35	24	8	12	0	44	10	8

Three nests of the wasps were collected before the emerging phase and glued to the tops of insect cages, which were left open when no experiments were being conducted, but otherwise were opened or closed according to the plan described below. Forty creeping thistle *Cirisium arvense* plants in a patch in a meadow were individually numbered, and early on 32 days in the summer of 2004, naturally occurring shield beetle larvae were removed and replaced by others placed visibly on randomly chosen thistle plants. Previous experiments had established that the larvae stay on the same leaf and that natural mortality is essentially non-existent, and video-recording showed that disappearance of larvae could be attributed almost entirely to predation by wasps. The 32 days were divided into eight blocks of four days, and treatments with different numbers of prey and open cages were applied with 0, 1, 2 or 3 open cages, and with $5, 10, \ldots, 35$ or 40 prey placed on randomly-chosen thistles. At the end of the day the plants were checked and the survival or not of each larva was noted; this is treated as a binary response. The number of individual wasps visiting the patch was estimated by extrapolation from the number visiting during the peak hour just after noon, which were caught, marked using dye, and released.

Figure 6.1 shows how the proportion of larvae eaten depends on the estimated number of wasps that visit the patch, on the plant number and on the number of larvae; it also shows how the number of wasps varies with the time, measured in days after June 1. Of the 720 larvae used overall, 635 were eaten, with only two days on which the proportions eaten were lower than 0.6; these were both days on which no cages were open and the

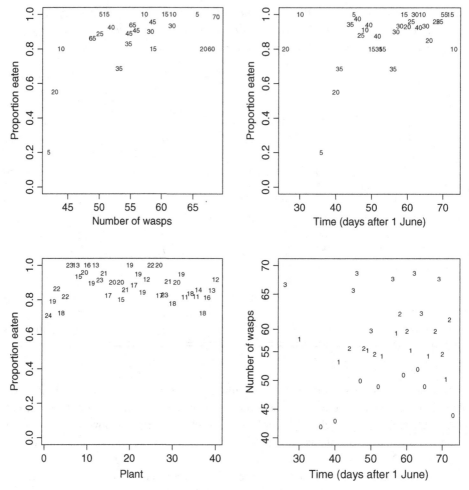

Figure 6.1 Summary plots for wasp data. The upper row and lower left panels show how the proportion of larvae eaten varies with the estimated number of wasps, the time, and the plant on which the larvae were placed. The plotting symbol gives the number of larvae contributing to each point. The lower right panel shows how the number of wasps depends on the time; the plotting symbol shows how many cages were open.

number of wasps present was low. If these two days are ignored, there seems to be little dependence of the numbers eaten on numbers of wasps, time or plant. The lower right panel shows an interesting pattern: when no cages are open, the numbers of wasps are low, and appear to show some seasonality; whereas when one or more cages is open, the numbers of wasps show no seasonality, and increases with the number of open cages. The evidence for seasonality when no cages are open is weak, however, because it depends essentially on three data points.

For a simple model, suppose that the probability that a single larva in the presence of n_w wasps survives to the end of a day is $\exp(-n_w^\gamma \xi)$, where $\xi > 0$. The quantity

$n_w^\gamma \xi$ can be interpreted as the daily attrition rate for an individual prey in the presence of n_w predators. The value of γ is of particular interest: $\gamma = 1$ corresponds to linear dependence of the rate on the number of wasps, as would occur if every wasp behaved independently of the others. The ecological question of interest is whether the predation efficiency of a group of wasps should increase more slowly than their numbers, corresponding to $0 \leq \gamma < 1$; this would indicate interference among predators. Values $\gamma > 1$ would suggest that attrition rate of larvae increases faster than linearly with the number of predators, indicating cooperation among them; this is ecologically unlikely.

This simple model postulates that prey are subject to independent attrition at the same rate, but in fact some larvae will be found more easily than others. We can model this by supposing that in the presence of n_w wasps the attrition rate for an arbitrarily chosen larva will be $n_w^\gamma \xi \varepsilon$, where ε is a positive random variable with density f, unit mean and variance ν. The probability of survival for one day is then

$$1 - p = \int_0^\infty e^{-n_w^\gamma \xi \varepsilon} f(\varepsilon) \, d\varepsilon.$$

If f is taken to be the gamma density, then

$$1 - p = \begin{cases} (1 + \nu n_w^\gamma \xi)^{-1/\nu}, & \nu > 0, \\ \exp(-n_w^\gamma \xi), & \nu = 0, \end{cases} \tag{6.1}$$

where the latter expression is the limit as $\nu \longrightarrow 0$. If $\log \xi = x^T \beta$ this corresponds to a generalized linear model with a binary response that indicates whether a larva has been eaten; this happens with probability p. The linear predictor for the generalized linear model is $\eta = x^T \beta + \gamma \log n_w$, and the complementary log–log and logit link functions

$$\eta = \log\{-\log(1 - p)\}, \quad \eta = \text{logit } p = \log\left(\frac{p}{1-p}\right),$$

arise when $\nu = 0$ and $\nu = 1$, respectively. Other link functions are given by taking (6.1) with $\nu > 0$.

Table 6.2 gives the reductions in the deviance – twice the difference between the log likelihood for the saturated model, which has one parameter per observation, and the fitted model – for a sequence of models in which the covariate vector x contains plant effects, effects of the number of prey, and $\log n_w$. When m larvae are placed on leaves we write

$$\log \xi_{jm} = \mu + \alpha_j + \eta_m + \gamma \log n_{w,m}, \quad j = 1, \ldots, 40, \quad m = 5, 10, \ldots, 40,$$

where j is the plant index and the estimated number of wasps present on this occasions is $n_{w,m}$. In fact data from just 33 plants may be used because seven plants had maximum likelihood estimates $\widehat{\alpha}_j = +\infty$: larvae placed on these plants were always eaten. When they are dropped, the total number of larvae drops from 720 to 594. We allow for the

Table 6.2 *Analysis of deviance for binary response models with logit and complementary log–log link functions fitted to the wasp data. Significance of an effect is gauged by comparing the change in deviance to a chi-squared distribution with the given degrees of freedom.*

Source	df	Logit		Complementary log–log	
		Deviance	Significance	Deviance	Significance
Plant	32	22.16	0.90	22.16	0.90
Number of prey	7	31.01	6.2×10^{-5}	30.89	6.5×10^{-5}
$\log n_w$	1	8.28	4.0×10^{-3}	9.40	2.2×10^{-3}
Residual	553	429.84		428.85	

total number of prey m in order to account for prey density, which plays an important role in classical predator–prey models. The plant effect is small for both models, and the number of prey is highly significant. The estimates $\hat{\eta}_m$ show no monotonic dependence on m, however, unlike what would be expected in a classical predator–prey model, so it is hard to interpret the effect of prey density. The deviance reductions given in Table 6.2 for inclusion of $\log n_w$ show that γ is significantly different from zero for both link functions. The likelihood ratio statistics for the hypothesis that $\gamma = 1$ are 3.55 and 0.88 for the logit and complementary log–log link functions respectively, with corresponding significance levels 0.06 and 0.35 when compared with the χ_1^2 distribution. The residual deviances suggest that the complementary log–log link function fits the data slightly better.

The logit link function corresponds to a logistic regression model, so the package cond can be used to investigate the effect of higher order correction on the fitted model. The 95% confidence intervals for γ based on the Wald pivot and on the likelihood root are $(0.890, 4.979)$ and $(0.924, 5.024)$, respectively. These intervals are first order, while the higher order interval based on the modified likelihood root is $(0.752, 4.674)$. This is shifted appreciably towards the origin, but all the intervals contain the value $\gamma = 1$ corresponding to no interference among wasps.

This conclusion depends strongly on the two left-most points in the upper left panel of Figure 6.1. When they are dropped, the deviance reductions due to $\log n_w$ are 0.47 and 1.55 for the logit and complementary log–log link functions, giving significance levels of 0.49 and 0.21 respectively when compared to a χ_1^2 distribution, and the confidence intervals for γ contain both $\gamma = 0$ and $\gamma = 1$: there is then no evidence that the probability that a larva is eaten depends on the number of wasps – as the bulk of the points in the panel would suggest.

When the complementary log–log link function is used the model is a curved exponential family, and (2.11) and (2.15) are needed for higher order inference. Here y_i follows a Bernoulli distribution whose success probability $p_i = p(x_i^{\mathsf{T}} \theta)$ depends on a linear combination of covariates x_i and parameters θ. The log likelihood can be written in the form (2.13) with $\alpha_i(\theta) = \text{logit } p(x_i^T \theta)$, so we take

$$V_i = \left.\frac{\partial \text{E}(y_i; \theta)}{\partial \theta^T}\right|_{\widehat{\theta}} = \left.\frac{\partial p(x_i^T \theta)}{\partial \theta^T}\right|_{\widehat{\theta}} = p'(x_i^T \widehat{\theta})x_i^T,$$

giving

$$\varphi(\theta)^T = \sum_{i=1}^n V_i \text{ logit } p(x_i^T \theta). \tag{6.2}$$

For the complementary log–log model we find that

$$V_i = \exp\{x_i^T \widehat{\theta} - \exp(x_i^T \widehat{\theta})\}x_i^T,$$

where $\widehat{\theta}$ is the overall maximum likelihood estimate. For the logistic regression model, $\text{logit } p(x_i^T \theta) = x_i^T \theta$, and (6.2) becomes

$$\varphi(\theta) = \theta^T(X^T W X),$$

where W is a diagonal matrix with elements $p'(x_i^T \widehat{\theta})$. In this case $\varphi(\theta)$ is an affine transformation of θ, the canonical parameter of the exponential family, so the inference is the same as when using the cond package.

The test of $\gamma = 1$ in the complementary log–log model and with effects of plants and number of prey yields Wald statistic $t = 0.934$ and likelihood root $r = 0.940$, with the 41 parameters present resulting in a substantial correction that yields modified likelihood root $r^* = 0.496$. None of these statistics suggests that $\gamma \neq 1$. The corresponding quantities without plant effects are 0.803, 0.806 and 0.697. Here the correction is smaller, though still perhaps surprisingly large for a situation in which nine parameters are fitted to 594 binary responses. Like the first order results, these depend strongly on the two observations mentioned above.

6.3 Grazing data

Figure 6.2 shows data from two experiments to assess the effect of cattle grazing on the regeneration of the wooded pastures of the Jura mountains. They were conducted in the summer of 2003 by Charlotte Vandenberghe and François Freléchoux of the Swiss Federal Institute for Forest, Snow and Landscape Research, and colleagues.

In the first experiment, sixteen square plots of side 8 metres were chosen randomly in each of two fields, and eight tree seedlings were planted at the mid-points and corners of each plot. The seedlings comprised four species of tree – fir, maple, beech and spruce – in pairs, one 20 cm and one 60 cm in height. Herds of cows were introduced into the two fields, and after a period it was observed whether or not each seedling had been grazed. This is a split-plot experiment in which the sub-plot design is a 4×2 factorial (species $\times$

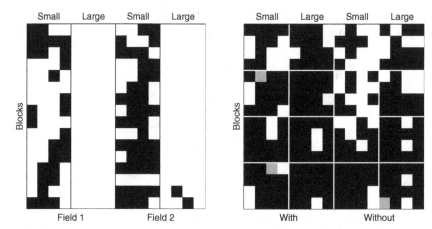

Figure 6.2 Graphical representation of data on cattle grazing. In the left panel, black denotes ungrazed and white denotes grazed seedlings, the vertical axis corresponds to the 16 blocks in each field, and species appear in order fir, maple, beech, spruce, within each set of four columns. In the right panel, black denotes alive and white denotes dead seedlings, and each set of four rows corresponds to the replicates in a single block; the species are in the same order as on the left. Three missing values are shown in grey.

height) and the response is binary (grazed or ungrazed) but with the two fields and hence the plots within them subject to different grazing pressures, owing to the different sizes. The total number of responses is

$$256 = 2 \text{ fields} \times 16 \text{ blocks} \times 4 \text{ species} \times 2 \text{ heights}.$$

The left-hand panel of Figure 6.2 shows that just two of the larger seedlings were ungrazed, and some difference between the fields – the higher grazing pressure in the smaller Field 1 results in more of the small seedlings being grazed, so the left half of the panel is whiter.

In a second experiment conducted to assess the effect of competition on seedling survival and growth, four areas of the larger field were closed to cattle. Half of each area was mowed, and four replications of the previous eight combinations of species and height were placed randomly into each half at the points of a regular grid, eventually yielding a total of a further 256 binary responses, indicating whether the seedlings were dead or alive after a specified period. The number of binary responses should again be

$$256 = 4 \text{ blocks} \times 2 \text{ treatments} \times 4 \text{ species} \times 2 \text{ heights} \times 4 \text{ replicates},$$

though in fact three are missing, giving 253 in total. The right-hand panel of Figure 6.2 shows that a fairly high proportion of the seedlings survived, with most of those dying being small seedlings without competition, that is, in mowed areas. Competition seems to have increased the survival rate, perhaps because mown areas provided less protection against dessication during the exceptionally hot summer of 2003.

We first consider the data from the first experiment, to which we fit a logistic regression model with effects of the 32 blocks, four species and height; field is aliased with the block effects. Not surprisingly the height effect is very large, the corresponding maximum likelihood estimate and standard error are $\widehat{\psi} = 6.37$ (1.28): there is overwhelming evidence that the larger seedlings are more likely to be grazed.

Higher order correction using the cond package directly fails because of convergence failure for values of ψ larger than $\widehat{\psi}$, and it is necessary to limit the range for which the model is refitted. Trial and error establish that the code in Code 6.1 works; note the use of the to option with cond. The data are contained in pref1, and pref1$x contains the indicator of large height. There is a very large adjustment, the adjusted maximum likelihood estimate $\widehat{\psi}_c$ and standard error being 4.76 (0.83). The confidence intervals based on the likelihood root and the modified likelihood root are respectively $(4.40, 9.19)$ and $(3.27, 6.14)$, corresponding to the anticipated huge odds ratio against survival for the larger plants.

The panels of Figure 6.3 compare the likelihood root $r(\psi)$ and its modified form $r^*(\psi)$ for the corresponding panels of Figure 6.2. Both show evidence of a strong effect of seedling height. Although the designs are perfectly balanced in both cases, there is a marked difference in the effect of higher order adjustment, presumably because there is very little information available for estimating the log odds ratio in the first data set: both information and nuisance parameter diagnostics are around eight times larger for the survival data in the left-hand panel than for the grazing preference data.

One question of interest is whether survival rates differ across species. For the data in the left-hand panel of Figure 6.2, the likelihood ratio statistic for the hypothesis of no species differences, after adjusting for the effects of block and height, is $w = 5.56$. When this is treated as χ_3^2, the resulting significance level is 0.135. Here the response distribution is discrete, so Bartlett adjustment of the likelihood ratio statistic is not guaranteed to give a theoretical improvement to the significance level, though simulation can be used to assess the quality of the χ_3^2 approximation. To perform such a simulation, we fit the simpler model, corresponding to the null hypothesis of interest, and use the fitted probabilities to generate binary responses. We then fit both the null and alternative models to the simulated data, and compute the corresponding likelihood ratio statistic, $w^\dagger$, say. This parametric bootstrap procedure is repeated enough times for the empirical properties of the resulting $w^\dagger$s to yield reliable estimates of the corresponding properties of w. In the present case the null model has effects of block and height, and we generated 5,000 bootstrap data sets,

Code 6.1 R code for approximate conditional inference for the grazing data.

```
fit <- glm( y ~ Block + Species + x, family = binomial,
            data = pref1 )
anova( fit, test = "Chisq" )
summary(fit)
fit.cond <- cond( fit, offset = x, to = 7 )
summary(fit.cond)
```

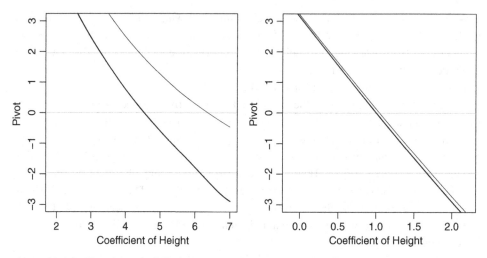

Figure 6.3 Likelihood root (solid) and modified likelihood root (heavy) for grazing preference data (left) and for seedling survival data (right). The horizontal lines are at $0, \pm 1.96$.

for which the average likelihood ratio statistic for the species effect is $\overline{w}^{\dagger} = 4.17$. Thus the adjusted likelihood ratio statistic equals $w \times 3/4.17 = 4.01$ with significance level 0.26 when treated as χ_3^2. This very large adjustment attenuates the already weak evidence for species differences in the data from the first experiment.

As mentioned in Section 4.1, exact tests are available for the logistic regression model. Suppose that under the null hypothesis the joint density of the independent binary responses $y_1, \ldots, y_n$ may be written as

$$\frac{\prod_{i=1}^{n} e^{x_i^{T} \theta y_i}}{\prod_{i=1}^{n} (1 + e^{x_i^{T} \theta})} = C(\theta)^{-1} \exp\left(\theta^{T} \sum_{i=1}^{n} x_i y_i\right),$$

where x_i is the $p \times 1$ vector of covariates associated to the ith response, and the normalising constant $C(\theta)$ does not depend on the response values. If the whole of θ is regarded as a nuisance parameter, then it may be eliminated by conditioning on the corresponding sufficient statistic $S = \sum x_i Y_i$. The resulting conditional density,

$$f(y \mid s) = \frac{1}{|\mathcal{Y}_s|}, \quad y \in \mathcal{Y}_s,$$

is uniform on the subset $\mathcal{Y}_s$ of the sample space

$$\mathcal{Y} = \{(y_1, \ldots, y_n) : y_i = 0, 1, i = 1, \ldots, n\}$$

for which $\sum x_i y_i = s$. The corresponding exact significance level is

$$p = \frac{|\{y : w(y) \geq w, y \in \mathcal{Y}_s\}|}{|\mathcal{Y}_s|},$$

where $w(y)$ denotes the likelihood ratio statistic for testing no species difference, computed for data y. In applications the enumeration of $\mathcal{Y}_s$ may be arduous, and a Monte Carlo estimate of p is often preferred; typically the Metropolis–Hastings algorithm is used for this purpose. The significance level is then exact apart from Monte Carlo error, and the test is called *Monte Carlo exact*.

In the present case the components of the x_i are indicator variables determining the block and height combination corresponding to each response; thus x_i has length $32 \times 2 = 64$ and the sum $\sum x_i y_i$ determines the number of ungrazed seedlings in each block-height combination. This is the number of black squares in each horizontal block of four squares shown in the left-hand panel of Figure 6.2, so the required conditional distribution is obtained by independent permutation of each of these quartets of observations; of course only mixed quartets contribute. The values of 5,000 likelihood ratio statistics thus obtained are shown in the left-hand panel of Figure 6.4. Their distribution is evidently very close to $c\chi_3^2$, with $c = 1.4$; the mean of the distribution is 4.20, very close to the value 4.17 obtained above using parametric simulation. The significance level is again adjusted to 0.26. Near the origin there are noticeable steps in the distribution of simulated values, because the number of possible data configurations under permutation limits the potential values of w under the conditional distribution.

The right-hand panel of Figure 6.4 shows the corresponding plot for the right-hand panel of Figure 6.2. In this case the likelihood ratio statistic for testing the hypothesis of no species differences, after adjusting for the effects of block, competition and height, is $w = 18.71$. When treated as χ_3^2 this yields a significance level of 0.0003. The simulated

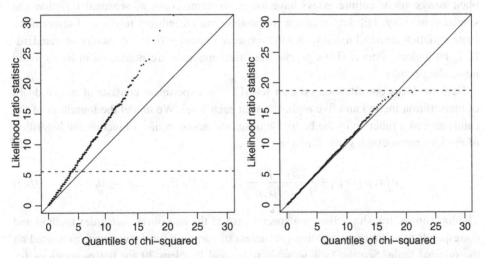

Figure 6.4 Empirical Bartlett adjustment for tests of species effect. Left: comparison of Monte Carlo exact conditional distribution for data in the left panel of Figure 6.2 with the χ_3^2 distribution. Right: corresponding comparison for the right panel of Figure 6.2. In both cases the likelihood ratio statistic is shown by the horizontal dashed line, and the diagonal line shows an exact match of the χ_3^2 and the empirical distribution.

distribution does not seem discrete near the origin, the mean of the replicate statistics is 3.10, and the adjusted likelihood ratio statistic is $w \times 3/3.10 = 18.11$ with corresponding significance level 0.0004. The conclusion of a strong difference among the survival rates for the different species is unchanged. The parameter estimates suggest that maple is appreciably more likely to survive than are the other species.

Analytical formulae for Bartlett adjustment factors have been published, but in cases such as this where considerable effort would be needed to program them, empirical Bartlett adjustment based on simulation is direct and straightforward.

6.4 Herbicide data

Seiden *et al.* (1998) present eight in vitro bioassays on the action of the sulfonylurea herbicides metsulfuron-methyl and chlorsulfuron on *Brassica napus L.* – also known as oilseed rape – a crop cultivated world-wide for the production of animal feed, vegetable oil for human consumption, and biodiesel. Six of the original data sets are available in the nlreg package as data frames C1, ..., C4, M2 and M4. The response variable area reports the callus area of colonies generated in tissue cultures and is expressed in mm^2. The variable dose represents the tested dose, measured in nmol/l. The data are plotted in Figure 6.5. The response is log-transformed to enhance readability.

In herbicide research, tissue culture assays are used in connection with screening for active compounds. The ultimate aim is to provide a means to assess whether a new crop can be safely planted in a field when there is herbicide left in the soil after earlier treatment. Obvious approaches to this are chemical investigation of the soil or whole plant assays; tissue culture assays have lower operating costs, no seasonal variation and similar sensitivity, but they measure the analyte via an indirect response. Furthermore, large variation among bioassays is often observed because of the difficulty of standardizing biological systems. Thus precise statistical methods are needed in order to avoid misleading results.

Here we focus our attention on data set M2. The experiment consists of $m = 8$ doses of metsulfuron-methyl and five replications at each level. We model the logarithm of the callus area as a function of the herbicide dose, and adopt as mean function the logarithm of the four-parameter logistic function

$$\mu(x; \beta) = \beta_1 + \frac{\beta_2 - \beta_1}{1 + (x/\beta_4)^{2\beta_3}}, \qquad x \geq 0, \ \beta_1, \ldots, \beta_4 \geq 0. \qquad (6.3)$$

The logarithmic transformation is a special case of the *transform-both-sides* method and corresponds to assuming a constant coefficient of variation for the response measured on the original scale. See the bibliographic notes and Problem 39 for the motivations for these choices. Model (6.3) results in an S-shaped dose-response function, where β_2 and β_1 represent the median callus area observed at the null and an infinite dose, respectively. The parameter β_4 is the ED_{50}, which represents the dose that gives a response half-way between β_1 and β_2, and β_3 determines the slope of the response curve around the ED_{50}.

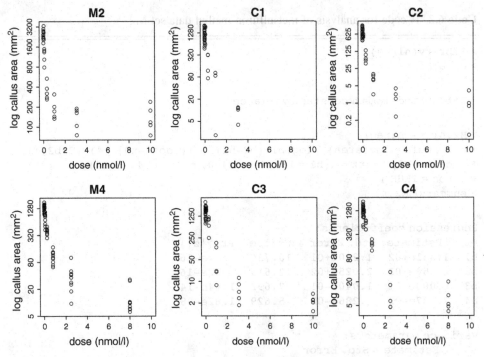

Figure 6.5 Six herbicide data sets from Seiden *et al.* (1998): effect of dose of metsulfuron-methyl (M2, M4) and chlorsulfuron (C1–C2) on tissue cultures of *Brassica Napus L.* (oilseed rape).

The model can be fitted by means of the `nlreg` fitting routine of the `nlreg` package. The corresponding R code is listed in Code 6.2.

As in the *Daphnia magna* example of Section 5.3, we may be interested in assessing whether there is a beneficial effect associated with low doses of the analyte. This can be done by fitting the extension of model (6.3), given on page 69, which includes a so-called hormesis effect. When this is fitted to the metsulfuron-methyl data set M2, the one-sided P-values for testing the significance of the hormesis parameter x_0 against the alternative that $x_0 > 0$ obtained from the Wald, the likelihood root, and from the modified likelihood root pivots (8.32) and (8.33) – the former using approximation (8.40) – are 0.0003, 0.048, 0.069 and 0.139. The first order Wald pivot is highly unreliable, because the finite sample distribution of the maximum likelihood estimator for x_0 is heavily skewed. The likelihood root statistic does somewhat better, although it is still not satisfactory. On the other hand, the two higher order pivots correct well for the non-normality of the distribution of $\widehat{x}_0$. According to the size of the information correction r_{INF}, this is especially the case for the r^* pivot (8.33). There is ample evidence against a growth effect at low herbicide doses. In fact this is a non-regular test, as the null parameter value lies on the edge of the parameter space, and the large-sample distribution of the likelihood ratio statistic is a 50–50 mixture of a χ_1^2 distribution and a point mass at zero; see the bibliographic notes.

Code 6.2 R code for analysis of metsulfuron-methyl data set M2.

```
> library(nlreg)
> data(M2)


# Fit TBS model with constant variance


> M2.nl <- nlreg(
+    formula = log(area) ~ log(b1+(b2-b1)/(1+((dose-x0)/b4)^(2*b3))),
+    data = M2, start = c(b1 = 130, b2 = 2400, b3 = 2, b4 = 0.05),
+    hoa = TRUE )
> summary( M2.nl )


Regression coefficients:
       Estimate  Std. Error   z value   Pr(>|z|)
b1    1.331e+02   1.240e+01    10.736    < 2e-16
b2    2.554e+03   2.025e+02    12.614    < 2e-16
b3    8.003e-01   1.040e-01     7.692   1.44e-14
b4    7.237e-02   1.286e-02     5.629   1.82e-08


Variance parameters:
       Estimate  Std. Error
logs     -2.590      0.2236


No interest parameter


Total number of observations: 40
Total number of parameters: 5
-2*Log Likelihood 9.904


# Graphical inspection of regression diagnostics


> nlreg.diag.plots( M2.nl )
 Make a plot selection (or 0 to exit)


1: plot: Summary
2: plot: Studentized residuals against fitted values
3: plot: r* residuals against fitted values
4: plot: Normal QQ-plot of studentized residuals
5: plot: Normal QQ-plot of r* residuals
6: plot: Cook statistic against h/(1-h)
7: plot: Global influence against h/(1-h)
8: plot: Cook statistic against observation number
9: plot: Influence measures against observation number
Selection:
```

Regression diagnostics are an important tool to assess the accuracy of the model fit, and are implemented in the `nlreg.diag.plots` routine. An application of them in Section 5.3 yielded the residuals plot shown in the left panel of Figure 5.3. When using the default setting `which = "all"`, the user can choose from a menu among eight different choices, and one summary plot; see Code 6.2. Figure 6.6 shows the output produced with option 1. The graphical device is split into four panels. The top left panel plots the studentized residuals against the fitted values, the top right panel the normal QQ-plot of the r^*-type residuals. See Problem 43 for how the second type of residuals is calculated. If the default setting `hoa = TRUE` is changed, higher order solutions such as the r^*-type residuals are not calculated. The top right panel of the plot is then replaced by a normal QQ-plot of the studentized residuals, and options 3 and 5 produce no output. The two panels in the bottom line are intended to highlight influential and high leverage points.

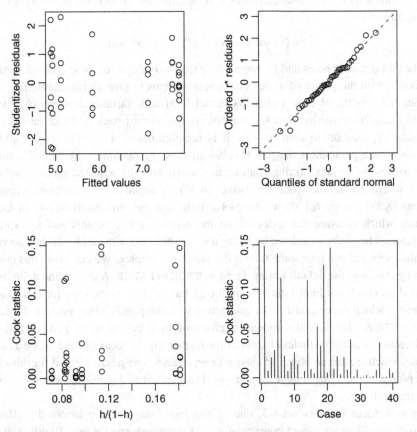

Figure 6.6 Metsulfuron-methyl data set M2 plots obtained with the `nlreg.diag. plots` routine of the `nlreg` package. Clockwise from top left: studentized residuals against fitted values; normal QQ-plot of r^*-type residuals; Cook's distance against observation number; Cook's distance against transformation of leverages.

They plot Cook's distance against the transformation $h_{ii}/(1 - h_{ii})$ of the leverages h_{ii} and against the observation number.

The residual plot for the unweighted least squares fit M2.nl shown in the top left panel of Figure 6.6 suggests that the error variance is not constant. A possible explanation is that the response level may vary according to how much chemical is absorbed by the plant rather than to how much chemical is present in the environment. If so, then the nominal dose level is a proxy measure for the absorption level of the individual tissue cultures. It is reasonable to assume a Berkson-type measurement error model, which gives rise to a variance function of the form

$$\sigma^2 w(x; \beta, \kappa, \gamma)^2 = \sigma^2 \left[1 + \kappa x^\gamma \left\{ \frac{d\mu(x; \beta)/dx}{\mu(x; \beta)} \right\}^2 \right], \quad \kappa, \gamma, \sigma^2 > 0, \qquad (6.4)$$

where κ, γ and σ^2 are non-negative variance parameters. We will refer to (6.4) as the *error-in-variables* (EIV) variance function. A simplified version is the *power-of-x* (POX) variance function

$$\sigma^2 w(x; \gamma)^2 = \sigma^2 (1 + x^\gamma)^2, \quad \gamma, \sigma^2 > 0. \qquad (6.5)$$

See the bibliographic notes and Problem 44 for the motivations for these choices. Refitting the model, with the power-of-x variance function seems to give an reasonable fit.

Code 6.3 lists the R code which fits model (6.3) with variance function (6.5) to the second metsulfuron-methyl data set available in the nlreg package (data set M4). The parameter β_1 was set to zero because it is not significant at the 5% level. Estimates of the variance parameters that are less biased than the maximum likelihood estimates are provided by the mpl fitting routine; this maximizes the adjusted profile likelihood (8.41), where the sample space derivatives (8.42) are replaced by Skovgaard's approximations (8.37). Figure 6.7 shows the global influence and two modifications of Cook's distance which measure the influence on the regression coefficients and the variance parameters. They were calculated by fitting a mean shift outlier model for each observation (Problem 43), and can take some time if the model is complex. We can avoid this computation by changing the default setting infl = TRUE to FALSE. A dotted line delimits any observation that has a high influence on the global fit. This is the case for observations 16 and 49, which were identified by interacting with the panels, made possible by setting iden = TRUE. By default a picture is produced by a call to nlreg.diag.plots. All diagnostics can nonetheless be saved for further use by means of the nlreg.diag routine, which creates an object of class nlreg.diag. The plot method for this object class produces the same output as nlreg.diag.plots. Figure 6.7 was obtained by invoking option 9.

We now focus on data set C3, one of the four assays run to assess the effect of chlorsulfuron. The experiment consists of $n = 51$ measurements for $m = 10$ different dose levels. The design is unbalanced, as the number of replicates per dose varies from 4 to 8. We again use the logarithm of the four-parameter logistic function (6.3) to model the logarithm of callus area as a function of the herbicide dose. The heteroscedasticity of the

Code 6.3 Model fit for the metsulfuron-methyl data set M4.

```
> library(nlreg)
> data(M4)

# Fit TBS model with POX variance (MLEs & MMPLEs)

> M4.pox.nl <-
+   nlreg( formula = log(area) ~ log(b2/(1+(dose/b4)^(2*b3)))),
+       weights = ~ (1+dose^g)^2, data = M4, hoa = TRUE,
+       start = c(b2 = 1500, b3 = 1, b4 = 0.05, g = 0.3) )

> M4.pox.mpl <- mpl( M4.pox.nl )
> summary( M4.pox.mpl )
Higher order method used: Skovgaard's r*

Variance parameters
          MMPLE      MLE    Std. Error
    g    0.5489    0.5496      0.1329
logs    -3.4377   -3.4803      0.1710

[... omitted ...]

Total number of observations: 72
Total number of parameters: 5
-2*Log Lmp 36.08s

# Use of diagnostic plots

> M4.pox.diag <- nlreg.diag( M4.pox.nl )
> plot( M4.pox.diag, which = 3 )
> plot( M4.pox.diag, which = 9, iden = TRUE )
****************************************************
Please Input a screen number (1,2 or 3)
0 will terminate the function
1: 1
Read 1 item
Interactive Identification for screen 1
left button = Identify, center button = Exit
```

error term is well modelled by the error-in-variables variance function (6.4). Code 6.4 gives the R code which fits this model. The chlorsulfuron data represent a challenging situation. To calculate higher order solutions we need the first and second derivatives of the mean and variance functions, and differentiating expression (6.4) twice with respect to the seven parameters imposes a non-negligible computational burden. As R is a purely numerical computing environment, its algebraic manipulation capabilities are limited. In

Code 6.4 R code for analysis of herbicide data set C3.

```
> library(nlreg)
> data(C3)

# Fit TBS model with EIV variance function

> C3.nl <-
+   nlreg( formula = log(area) ~ log(b1+(b2-b1)/(1+(dose/b4)^(2*b3))),
+         weights = ~ 1 + k * dose^g *
+                     ( (b2-b1)*(2*b3)*dose^(2*b3-1)/
+                       (1+(dose/b4)^(2*b3))^2/b4^(2*b3)/
+                       (b1+(b2-b1)/(1+(dose/b4)^(2*b3))) )^2 ,
+         start = c(b1=2.2, b2=1700, b3=2.8, b4=0.28, g=2.7, k=1),
+         data = C3, hoa = TRUE, trace = TRUE,
+         control = list(x.tol=1e-12, rel.tol=1e-12, step.min=1e-12))
> summary( C3.nl )

Regression coefficients:
      Estimate  Std. Error  z value  Pr(>|z|)
b1   2.206e+00   4.149e-01    5.317  1.05e-07
b2   1.664e+03   1.175e+02   14.154   < 2e-16
b3   1.420e+00   1.800e-01    7.893  2.95e-15
b4   2.751e-01   4.522e-02    6.085  1.17e-09

Variance parameters:
        Estimate  Std. Error
g          2.605      0.7931
k          1.009      0.5796
logs      -1.888      0.2341

No interest parameter

Total number of observations: 51
Total number of parameters: 7
-2*Log Likelihood 70.21
```

these situations, it is recommended that all symbolic derivatives be stored as part of the fitted model object to save execution time.

Similarly, model (6.3) can be fitted to the remaining three chlorsulfuron data sets. We use the power-of-x variance function for data sets C1 and C4, while data set C2 requires the error-in-variables variance function. The parameter β_1 turns out to be statistically not significant at the 5% level in all three model fits. For the sake of illustration, Figure 6.8 shows the profile plots and approximate bivariate contour plots for all parameters in the model fitted to data set C2. The sample space derivatives required in the calculation of the higher order pivots are replaced by Skovgaard's approximations (8.37).

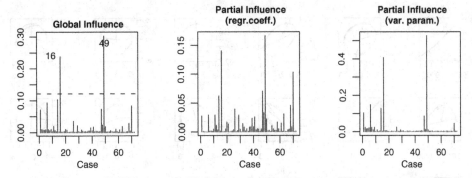

Figure 6.7 Metsulfuron-methyl data set M4: influence measures, obtained from the routine `nlreg.diag. plots` of the `nlreg` package. Left panel: global influence measure. Middle and right panels: Cook's distances measuring the influence on the regression coefficients and on the variance parameters. The figures 16 and 49 in the left panel identify influential observations.

Among the objectives of Seiden *et al.* (1998) were the assessment of the theoretical 'limit of detection' of the two herbicides and the evaluation of the reproducibility of the results thus obtained across bioassays. In general terms, a limit of detection is defined as the highest dose of the analyte which has no significant effect on the tested organism. There are various quantitative translations of this definition according to the different branches of the biological sciences where they are used. Reproducibility of the assays is usually measured in terms of the coefficient of variation of a central value such as the ED_{50}.

The *statistical-no-effect* dose (SNED) relies upon the concept of confidence bands for the response. Mathematically, it is defined as the maximum dose level at which the response does not differ from the level observed for the controls with 95% confidence. This corresponds to the maximum dose

$$x_{\text{sned}} = \max\{x \mid L_\alpha(x, \widehat{\beta}) \le \mu(0; \widehat{\beta}) \le U_\alpha(x, \widehat{\beta})\}$$

at which the response $\mu(0; \widehat{\beta})$ for zero dose is included in the 95% confidence interval $[L_\alpha(x, \widehat{\beta}) , U_\alpha(x, \widehat{\beta})]$ for the mean function $\mu(x; \beta)$. Operationally, it can be obtained as illustrated in the left panel of Figure 6.9, where the heavy line represents the fitted response $\mu(x; \widehat{\beta})$ and the two embracing solid lines are respectively the upper and lower 95% confidence bands. The right panel of Figure 6.9 shows, from left to right, the SNEDs obtained from the first order Wald (black) and likelihood root (dark grey) pivots and from the two higher order pivots (8.32) and (8.33) (grey and light grey) for the six data sets considered here. The four methods yield similar results for a given assay, although the values obtained from the higher order pivots tend to be slightly larger. The same cannot be said if we compare the values among paired assays. The SNEDs vary in the range 0.0116–0.0224 nmol/l for metsulfuron and 0.0375–0.174 nmol/l for chlorsulfuron. The squared coefficient of variation for the latter case varies between 41.0% and 52.6%, which indicates that standardization of assays requires consideration.

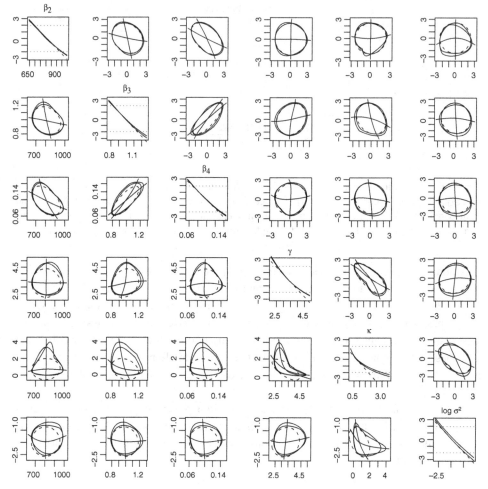

Figure 6.8 Chlorsulfuron data set C2: profile plots and approximate bivariate contour plots obtained with the `contour` method of the `nlreg` package (——, w/r; ——, w^*/r^*; - - -, Wald).

An alternative to the use of the SNED is to calculate the dose ED_p which decreases the response by $p \times 100\%$. These values are called *effect* doses. Columns 2–4 of Table 6.3 report the ED_5, ED_{10} and ED_{50} for the six herbicide data sets. The ED_{50} corresponds to β_4, while a general effect dose is characterized by the relationship

$$\left(\frac{1-p}{p}\right)^{1/(2\beta_3)} ED_p = ED_{50}.$$

Columns 5–7 characterize the fitted model. Chlorsulfuron gives slightly steeper curves, while metsulfuron gives slightly smaller effect doses. The squared coefficients of variation

Table 6.3 *Effect doses (ED_p) in nmol/l for the six herbicide data sets.*
Columns 2–4: 5%, 10% and 50% effect doses in nmol/l for the six herbicide
data sets. Columns 5–7: characteristics of the fitted models and maximum
likelihood estimates of the shape parameters.

Data set	ED_5	ED_{10}	ED_{50}	Variance function	β_1	$\widehat{\beta}_3$
M2	0.0144	0.0220	0.0774	POX	present	0.80
M4	0.00704	0.0126	0.0982	POX	absent	0.57
C1	0.0241	0.0370	0.1330	POX	absent	0.85
C2	0.0222	0.0324	0.0973	EIV	absent	0.98
C3	0.0983	0.1280	0.2760	EIV	present	1.42
C4	0.0435	0.0677	0.2380	POX	absent	0.87

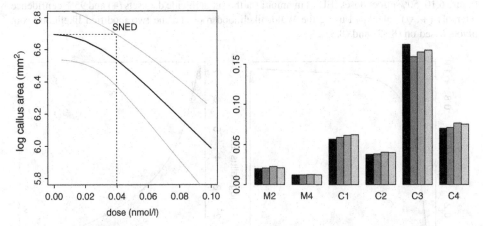

Figure 6.9 Statistical-no-effect doses (SNEDs) in nmol/l for the six herbicide data sets. Left panel: the SNED is the largest close at which 95% confidence limits for the mean response include the mean response at zero close. Right panel: barcharts of the SNEDs obtained from the Wald (black), likelihood root (dark grey) and the two modified likelihood root pivots (8.32) and (8.33) (grey and light grey).

are 57.0%, 44.0% and 20.7% for the ED_5, ED_{10} and ED_{50}, respectively. Figure 6.10 reports the confidence intervals for the EC_{50} for the six data sets.

In analogy to our discussion of the *Daphnia magna* example of Section 5.3, we may calculate the 50% '*lethal*' dose LD_{50}, that is, the dose which reduces the callus area below a given value in half the population. This is closely connected with the effect dose EC_p, though these two measures differ from the toxicological viewpoint. In the first case a whole population is observed, and the average decrease in the response is recorded. In the second case, the response of individual subjects is observed. Thanks to the transform-both-sides method, it is straightforward to obtain the LD_{50} values for a given response level from the ED_p estimates: the LD_{50} at level y_0 coincides with the ED_p for which $\mu(ED_p; \beta) = y_0$. Figure 6.11 shows the ED_p values and the LD_{50} values for data set C4

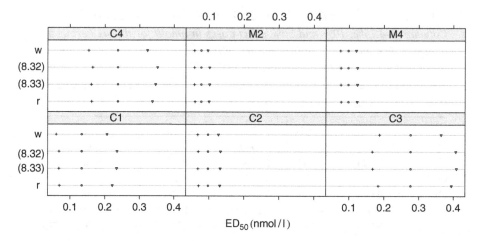

Figure 6.10 50% effect doses (ED_{50}) in nmol/l for the six herbicide data sets (+) and 95% confidence intervals (+, ▽) calculated using the Wald, likelihood root and the two modified likelihood root pivots based on (8.32) and (8.33).

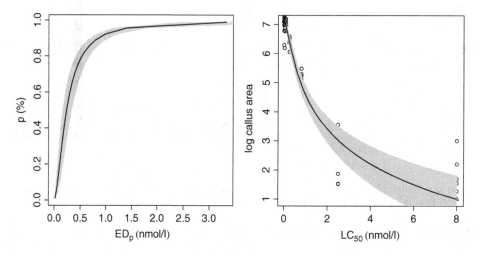

Figure 6.11 Effect doses and 50% lethal doses in nmol/l for data set C4 together with the corresponding 95% pointwise confidence bands (shaded) calculated using the likelihood root.

together with the 95% pointwise confidence bands obtained from the likelihood root. The doses which entail a 5% and 10% decrease of the callus area in half the population are respectively $LD_5 = 0.134$ and $LD_{10} = 0.239$ nmol/l.

Bibliographic notes

Ridout *et al.* (2006) discuss analysis of an experiment with bees that has some features of the wasp experiment in Section 6.2. Analysis of binary data with non-canonical link functions is described in most accounts of generalized linear models; see for example McCullagh and Nelder (1989, Chapter 4). The complementary log–log link arises in the fitting of serial dilution assays; see Ridout (1994) and Problem 42. The construction of V and φ for use with discrete data is taken from Davison *et al.* (2006).

The potential failure of Bartlett adjustment with discrete data was noted in Frydenberg and Jensen (1989). Estimating the Bartlett correction by bootstrapping was suggested in Bickel and Ghosh (1990) and used in Bellio (2000); we use it also in Section 7.5. The use of the conditional distribution to assess goodness-of-fit is discussed in Cox and Hinkley (1974, Chapter 3).

Parametric Monte Carlo tests are discussed in Section 4.2 of Davison and Hinkley (1997), including a brief account of the application of Markov chain simulation in this context. See also Besag and Clifford (1989, 1991), Forster *et al.* (1996) and Smith *et al.* (1996).

The use of the four-parameter logistic function and of the transform-both-sides method for the herbicide bioassay data was suggested in Seiden *et al.* (1998). The transform-both-sides method is described in Carroll and Ruppert (1988). Two of the eight original assays were not satisfactorily described by this model. The remaining six were further analyzed in Bellio *et al.* (2000), who introduced the error-in-variables and power-of-x variance functions (Rudemo *et al.*, 1989) and discussed the use of higher order asymptotics. Growth stimulation at low doses can be accounted for through the model proposed by Brain and Cousens (1989). The statistical-no-effect dose was introduced by Chèvre *et al.* (2002).

7

Further topics

7.1 Introduction

In this chapter we illustrate the breadth of application of higher order asymptotics by presenting a variety of examples, most of which have appeared in the published literature. In contrast to the earlier chapters, the emphasis is on the methods for higher order approximation, with the data treated as mainly illustrative. Section 7.2 outlines a problem of calibration in normal linear regression, and the two succeeding sections outline higher order approximation for a variance components setting and for dependent data, respectively. Sections 7.5 and 7.6 concern a problem of gamma regression; we compare Bartlett correction to Skovgaard's multivariate adjustment to the likelihood ratio statistic, and indicate the use of Laplace approximation for Bayes inference. In Section 7.7 we consider if it is worthwhile to apply higher order approximation to partial likelihood. The final section concerns use of a constructed exponential family to find the distribution of interest for a randomisation test.

7.2 Calibration

Table 7.1 shows measurements of the concentration of an enzyme in human blood plasma. The true concentration x is obtained using an accurate but expensive laboratory method, and the measured concentration y is obtained by a faster and less expensive automatic method. The goal is to use the observed data pairs to estimate values of the true concentration based on further measurements using the less expensive method. This is an example of a calibration problem: we have a model for $E(y \mid x)$ that depends on some unknown parameters, and use a sample of pairs $(x_1, y_1), \ldots, (x_n, y_n)$ to estimate these parameters. The same model is then used with a further sample $y_{0j}, j = 1, \ldots, m_0$, to estimate the true concentration in this new sample.

The simplest calibration problem is a linear regression, in which case we have

$$y_i = \beta_0 + \beta_1 x_i + \varepsilon_i, \quad i = 1, \ldots, n, \quad \varepsilon_i \sim N(0, \sigma^2), \tag{7.1}$$

$$y_{0j} = \beta_0 + \beta_1 \psi + \varepsilon_{0j}, \quad j = 1, \ldots, m_0, \quad \varepsilon_{0j} \sim N(0, \sigma^2), \tag{7.2}$$

Table 7.1 *Data for linear calibration (Aitchison and Dunsmore, 1975, p. 183); x is the true, expensive, measurement and y is the less expensive measurement.*

x	3.0	3.4	3.8	4.2	4.6	5.0	5.4	5.8	6.2
y	2.3	2.6	3.0	3.2	3.7	3.9	4.2	4.6	4.9
	2.4	2.8	3.0	3.3	3.7	4.0	4.2	4.7	5.0
	2.5	2.8	3.1	3.4	3.7	4.1	4.3	4.8	5.2

where we assume that the ε_i and ε_{0j} are mutually independent. In the calibration literature ψ is usually denoted x_0; we use ψ to emphasize that it is the parameter of interest. The full parameter is $\theta = (\beta_0, \beta_1, \sigma, \psi)$, and the nuisance parameter is $\lambda = (\beta_0, \beta_1, \sigma)$.

In this problem the exact pivot

$$t_\psi = \frac{\bar{y} - \widehat{\beta}_0 - \widehat{\beta}_1 \psi}{s \left\{ \dfrac{1}{m_0} + \dfrac{1}{n} + \dfrac{(\bar{x} - \psi)^2}{S_{xx}} \right\}^{1/2}}$$

has the Student t_{n+m_0-3} distribution; here

$$\widehat{\beta}_0 = \bar{y} - \widehat{\beta}_1 \bar{x}, \qquad \widehat{\beta}_1 = S_{xy}/S_{xx},$$

$$S_{xy} = \sum_{i=1}^n (y_i - \bar{y})(x_i - \bar{x}), \qquad S_{xx} = \sum_{i=1}^n (x_i - \bar{x})^2,$$

$$s^2 = \frac{1}{n + m_0 - 3} \left\{ \sum_{i=1}^n (y_i - \widehat{\beta}_0 - \widehat{\beta}_1 x_i)^2 + \sum_{j=1}^{m_0} (y_{0j} - \bar{y}_0)^2 \right\}.$$

In the computation of the full maximum likelihood estimate we have

$$\bar{y}_0 = \widehat{\beta}_0 + \widehat{\beta}_1 \widehat{\psi},$$

showing that $\widehat{\beta}_0$ and $\widehat{\beta}_1$ based on $(y_1, \ldots, y_n, y_{01}, \ldots, y_{0m_0})$ equal those based on $(y_1, \ldots, y_n)$. The maximum likelihood estimate of σ^2 is

$$\widehat{\sigma}^2 = \frac{n + m_0 - 3}{n + m_0} s^2.$$

We will illustrate the use of the third order approximation $\Phi(r^*)$ with q defined by (8.33), and compare this to some related approximations from the literature. Our discussion is taken from Bellio (2003) and Eno (1999).

The model determined by (7.1) and (7.2) is a full exponential family with sufficient statistic

$$\left(\sum_{i=1}^n y_i, \sum_{i=1}^n y_i x_i, \sum_{i=1}^n y_i^2 + \sum_{j=1}^{m_0} y_{0j}^2, \sum_{j=1}^{m_0} y_{0j} \right)^{\mathrm{T}}$$

and canonical parameter

$$\varphi(\theta) = \left(\frac{\beta_0}{\sigma^2}, \frac{\beta_1}{\sigma^2}, -\frac{1}{2\sigma^2}, \frac{\mu_0(\psi)}{\sigma^2} \right)^{\mathrm{T}},$$

where $\mu_0(\psi) = \beta_0 + \beta_1 \psi$.

In addition to the likelihood root, r, we will need expressions for the constrained maximum likelihood estimator $\widehat{\theta}_\psi$, information determinant $|j_{\lambda\lambda}|$, and matrix of partial derivatives $\varphi_\theta(\theta) = \partial\varphi/\partial\theta^{\mathrm{T}}$. The expressions for the constrained maximum likelihood estimates are simpler if, following Bellio (2003), we define

$$I(\psi_1, \psi_2) = (n + m_0)S_{xx} + nm_0(\overline{x} - \psi_1)(\overline{x} - \psi_2).$$

We then have

$$\widehat{\beta}_{0\psi} = \widehat{\beta}_0 + \frac{\widehat{\beta}_1}{I(\psi, \psi)} m_0 \{S_{xx} + n(\overline{x} - \psi)\},$$

$$\widehat{\beta}_{1\psi} = \widehat{\beta}_1 \frac{I(\psi, \widehat{\psi})}{I(\psi, \psi)},$$

$$\widehat{\sigma}_\psi^2 = \widehat{\sigma}^2 + \frac{\widehat{\beta}_1^2 (\widehat{\psi} - \psi)^2 nm_0 S_{xx}}{(n + m_0)I(\psi, \psi)},$$

$$|j_{\lambda\lambda}(\widehat{\psi}, \widehat{\lambda})| = \frac{2(n + m_0)}{\widehat{\sigma}^6} I(\widehat{\psi}, \widehat{\psi}),$$

$$|j_{\lambda\lambda}(\psi, \widehat{\lambda}_\psi)| = \frac{2(n + m_0)}{\widehat{\sigma}_\psi^6} I(\psi, \psi).$$

The other quantities needed to compute q are obtained from the matrix

$$\varphi_\theta(\theta) = \frac{1}{\sigma^2} \begin{pmatrix} 1 & 0 & -2\beta_0/\sigma & 0 \\ 0 & 1 & -2\beta_1/\sigma & 0 \\ 0 & 0 & 1/\sigma & 0 \\ 1 & \psi & 2(\beta_0 + \beta_1\psi)/\sigma & \beta_1 \end{pmatrix},$$

whose first three columns will be used to construct the sub-matrix $\varphi_\lambda(\widehat{\theta}_\psi)$.

Table 7.2 gives 90% confidence intervals for the true concentration ψ, both for one further observation at $y_0 = 5.2$ and for one further observation at $y_0 = 3.7$. We constructed the intervals based on r^* using the expression for q_2 given at (8.33), using a general program to compute this from φ. Bellio (2003) gives an explicit expression for the Barndorff-Nielsen version q_1 given at (8.32), which in this case is identical to q_2, since no ancillary statistic is involved. Eno (1999, Chapter 3) compares several posterior intervals, for various choices of various non-informative priors. We include here the solution based on the *reference* prior; it gives results nearly identical to those based on r^*. Priors for which the posterior intervals are confidence intervals are called matching priors. Reference

Table 7.2 *Linear calibration using data from Table 7.1. Comparison of 90% confidence (or posterior) interval estimates for ψ, using first order and higher order approximations.*

	$y_0 = 3.7$		$y_0 = 5.2$	
Normal approximation to Wald pivot	4.439	4.806	6.272	6.660
Normal approximation to likelihood root	4.434	4.811	6.259	6.667
Normal approximation to r^* using q_2 (8.33)	4.421	4.824	6.255	6.681
Exact solution based on t_{26} pivot	4.418	4.828	6.252	6.685
Bayes posterior using reference prior, from Eno (1999)	4.421	4.825	6.255	6.682

priors are not guaranteed to be matching priors, but they are in many cases, and Eno (1999, Chapter 4) shows that for the linear calibration model the reference priors are indeed matching priors. A simulation study reported in Bellio (2000, Chapter 3) verifies that the coverage of confidence intervals based on r^* is essentially exact.

7.3 Variance components

Table 7.3 contains data on tests of voltage regulators for automobiles, from Cox and Snell (1981, Example S). A regulator selected randomly from the production line was adjusted at one of a number of setting stations, and then tested at each of four testing stations, giving the data in the table. A random sample of ten of the possible setting stations took part. The measured voltages for the 64 regulators involved are mostly quite consistent across testing stations, with the exception of regulators B_2 and J_2; in the absence of further information we include these in the analysis below. Interest here lies primarily in the random variation among setting stations and regulators, and less in any systematic variation owing to the testing stations. One quantity of interest was the proportion of regulators whose notional 'true' voltage would lie between 15.8 and 16.4 volts if the population mean were kept equal to 16.1 volts.

Table 7.4 shows the analysis of variance for these data. The interaction between setting stations and testing stations seems to be accountable by random error, but there is appreciable variation between setting stations, testing stations, and regulators within setting stations. The corresponding linear model for the voltage for regulator r adjusted at setting station s and passed to testing station t, is

$$y_{srt} = \mu_t + \alpha_s + \eta_{sr} + \varepsilon_{srt}, \quad s = 1, \ldots, 10, \ r = 1, \ldots, n_s, \ t = 1, \ldots, 4, \qquad (7.3)$$

where the α_s, η_{sr} and ε_{srt} are mutually independent normal random variables with zero means and variances σ_α^2, σ_η^2 and σ^2, and the μ_t are regarded as fixed unknown parameters corresponding to the four testing stations. Here the random choice of regulators and

Table 7.3 *Regulator voltages (Cox and Snell, 1981, p. 140)* .

Setting station	Regulator number	Testing station				Setting station	Regulator number	Testing station			
		1	2	3	4			1	2	3	4
A	1	16.5	16.5	16.6	16.6	F	1	16.1	16.0	16.0	16.2
	2	15.8	16.7	16.2	16.3		2	16.5	16.1	16.5	16.7
	3	16.2	16.5	15.8	16.1		3	16.2	17.0	16.4	16.7
	4	16.3	16.5	16.3	16.6		4	15.8	16.1	16.2	16.2
	5	16.2	16.1	16.3	16.5		5	16.2	16.1	16.4	16.2
	6	16.9	17.0	17.0	17.0		6	16.0	16.2	16.2	16.1
	7	16.0	16.2	16.0	16.0		11	16.0	16.0	16.1	16.0
	11	16.0	16.0	16.1	16.0						
						G	1	15.5	15.5	15.3	15.6
							2	16.0	15.6	15.7	16.2
B	1	16.0	16.1	16.0	16.1		3	16.0	16.4	16.2	16.2
	2	15.4	16.4	16.8	16.7		4	15.8	16.5	16.2	16.2
	3	16.1	16.4	16.3	16.3		5	15.9	16.1	15.9	16.0
	4	15.9	16.1	16.0	16.0		6	15.9	16.1	15.8	15.7
							7	16.0	16.4	16.0	16.0
							12	16.1	16.2	16.2	16.1
C	1	16.0	16.0	15.9	16.3						
	2	15.8	16.0	16.3	16.0	H	1	15.5	15.6	15.4	15.8
	3	15.7	16.2	15.3	15.8		2	15.8	16.2	16.0	16.2
	4	16.2	16.4	16.4	16.6		3	16.2	15.4	16.1	16.3
	5	16.0	16.1	16.0	15.9		4	16.1	16.2	16.0	16.1
	6	16.1	16.1	16.1	16.1		5	16.1	16.2	16.3	16.2
	10	16.1	16.0	16.1	16.0		10	16.1	16.1	16.0	16.1
						J	1	16.2	16.1	15.8	16.0
D	1	16.1	16.0	16.0	16.1		2	16.2	15.3	17.8	16.3
	2	16.0	15.9	16.2	16.0		3	16.4	16.7	16.5	16.5
	3	15.7	15.8	15.7	15.7		4	16.2	16.5	16.1	16.1
	4	15.6	16.4	16.1	16.2		5	16.1	16.4	16.1	16.3
	5	16.0	16.2	16.1	16.1		10	16.4	16.3	16.4	16.4
	6	15.7	15.7	15.7	15.7						
	11	16.1	16.1	16.1	16.0	K	1	15.9	16.0	15.8	16.1
							2	15.8	15.7	16.7	16.0
							3	16.2	16.2	16.2	16.3
E	1	15.9	16.0	16.0	16.5		4	16.2	16.3	15.9	16.3
	2	16.1	16.3	16.0	16.0		5	16.0	16.0	16.0	16.0
	3	16.0	16.2	16.0	16.1		6	16.0	16.4	16.2	16.2
	4	16.3	16.5	16.4	16.4		11	16.0	16.1	16.0	16.1

Table 7.4 *Analysis of variance for regulator voltages.*

	Sum of squares	df	Mean square	*F* value
Setting stations (*SS*)	4.4191	9	0.4910	9.08
Testing stations (*TS*)	0.7845	3	0.2615	4.84
SS × *TS*	0.9041	27	0.0335	0.62
Regulators within *SS*	9.4808	54	0.1756	3.25
Residual	8.7614	162	0.0541	

setting stations implies that these should be regarded as selected at random from suitable populations. As mentioned above, interest focuses on the variances, the μ_t being essential to successful modelling but of secondary interest.

Under model (7.3) and owing to the balanced structure of the data, the quantities

$$y_{srt} - \bar{y}_{sr\cdot} - \bar{y}_{s\cdot t} + \bar{y}_{s\cdot\cdot}, \quad \bar{y}_{sr\cdot} - \bar{y}_{s\cdot\cdot}, \quad \bar{y}_{s\cdot t} - \bar{y}_{s\cdot\cdot}, \quad \bar{y}_{s\cdot\cdot}, \quad r = 1, \ldots, n_s, \quad t = 1, \ldots, 4,$$

are independent normal variables for each setting station s; as usual, the overbar and replacement of a subscript by a dot indicates averaging over that subscript. Thus the residual sum of squares, the sum of squares between regulators within setting stations, and the sum of squares between testing stations,

$$SS_s = \sum_{r,t}(y_{srt} - \bar{y}_{sr\cdot} - \bar{y}_{s\cdot t} + \bar{y}_{s\cdot\cdot})^2,$$

$$SS_s^R = \sum_{r,t}(\bar{y}_{sr\cdot} - \bar{y}_{s\cdot\cdot})^2,$$

$$SS_s^T = \sum_{r,t}(\bar{y}_{s\cdot t} - \bar{y}_{s\cdot\cdot})^2,$$

are independent. Their distributions are respectively

$$\sigma^2 \chi^2_{3(n_s-1)}, \quad (\sigma^2 + 4\sigma_\eta^2)\chi^2_{n_s-1}, \quad \sigma^2 \chi^2_3(\delta),$$

the last being non-central chi-squared with non-centrality parameter $\delta = \sum_t(\mu_t - \bar{\mu}_\cdot)^2$. The average voltage $\bar{y}_{s\cdot\cdot}$ for setting station s has a normal distribution with mean $\bar{\mu}_\cdot$ and variance $\sigma_\alpha^2 + \sigma_\eta^2/n_s + \sigma^2/(4n_s)$. The distribution of the sum of squares for testing stations SS_s^T depends only on the μ_t and σ^2, and so contributes only indirectly to inference on the other variances. As there are 162 degrees of freedom available to estimate σ^2 from the residual sum of squares, the gain from inclusion of SS_s^T will be slight, and we base a marginal log likelihood on the other summary statistics.

Apart from constants, the contribution to the overall log marginal likelihood from the *s*th setting station is therefore

$$\ell_s(\theta) = -\frac{1}{2}\left[\frac{SS_s}{\sigma^2} + 3(n_s-1)\log\sigma^2 + \frac{SS_s^R}{\sigma^2+4\sigma_\eta^2} + (n_s-1)\log(\sigma^2+4\sigma_\eta^2) \right.$$

$$\left. + \frac{(\bar{y}_{s\cdot\cdot}-\bar{\mu}_\cdot)^2}{\sigma_\alpha^2+\sigma_\eta^2/n_s+\sigma^2/(4n_s)} + \log\left\{\sigma_\alpha^2+\sigma_\eta^2/n_s+\sigma^2/(4n_s)\right\} \right].$$

The overall log marginal likelihood $\sum_s \ell_s$ depends on the residual sum of squares $SS = \sum_s SS_s$, the sum of squares for regulators within setting stations $SS^R = \sum_s SS_s^R$, and the average voltages $\bar{y}_{s\cdot\cdot}$ for each of the setting stations, and these are mutually independent. The argument of Section 2.4 can therefore be applied to the twelve log marginal likelihood contributions.

Use of expression (2.11) for inference depends on computation of the local parametrization $\varphi(\theta)$ defined at (2.12), and this requires the quantities V_i. In this continuous model we can use expression (8.19) with pivots

$$z_s = \frac{\bar{y}_{s\cdot\cdot}-\bar{\mu}_\cdot}{\left\{\sigma_\alpha^2+\sigma_\eta^2/n_s+\sigma^2/(4n_s)\right\}^{1/2}}, \quad s = 1,\ldots,10,$$

based on the averages, and $z_{11} = SS/\sigma^2$ and $z_{12} = SS^R/(\sigma^2+4\sigma_\eta^2)$ based on the sums of squares. With $\theta^T = (\sigma_\alpha^2, \sigma_\eta^2, \sigma^2, \bar{\mu}_\cdot)$ this yields the following vectors:

$$V_s = \frac{1}{2c_s}(\hat{z}_s, \hat{z}_s/n_s, \hat{z}_s/(4n_s), c_s), \quad s = 1,\ldots,10,$$

$$V_{11} = (0, 0, SS/\hat{\sigma}^2, 0),$$

$$V_{12} = \frac{SS^R}{\hat{\sigma}^2+4\hat{\sigma}_\eta^2}(0, 4, 1, 0),$$

where $c_s = \{\hat{\sigma}_\alpha^2+\hat{\sigma}_\eta^2/n_s+\hat{\sigma}^2/(4n_s)\}^{1/2}$ and $\hat{z}_s = (\bar{y}_{s\cdot\cdot}-\hat{\mu}_\cdot)/c_s$. Further algebraic simplification of these expressions is possible. The corresponding likelihood derivatives with respect to the data are

$$\frac{\partial\ell}{\partial\bar{y}_{s\cdot\cdot}} = -\frac{\bar{y}_{s\cdot\cdot}-\bar{\mu}_\cdot}{\sigma_\alpha^2+\sigma_\eta^2/n_s+\sigma^2/(4n_s)}, \quad s = 1,\ldots,10,$$

$$\frac{\partial\ell}{\partial SS} = -\frac{1}{2\sigma^2},$$

$$\frac{\partial\ell}{\partial SS^R} = -\frac{1}{2(\sigma^2+4\sigma_\eta^2)},$$

which are weighted by the Vs to produce $\varphi(\theta)$. All that is needed to obtain the local parametrizations for the other parameters, and for functions of them, is to rewrite the

Code 7.1 R code for higher order inference for σ_α^2 for the voltage regulator data.

```
n <- c( 8, 4, 7, 7, 4, 7, 8, 6, 6, 7 )
ybar <- c( 16.3375,  16.1625,  16.05357, 15.96429, 16.16875,
           16.22143, 15.97812, 16,       16.29583, 16.09286 )
volty <- list( ss = 8.7614, ss.df = 162, ssr = 9.4808, ssr.df = 54,
               ybar = ybar, n = n )

nlogL <- function(psi, lam, d) {
# compute negative log likelihood
  sig2a <- exp(psi)
  sig2e <- exp(lam[1])
  sig2 <- exp(lam[2])
  mu <- lam[3]
  L <- d$ss/sig2 + log(sig2)*d$ss.df +
       d$ssr/(sig2 + 4*sig2e) + log(sig2 + 4*sig2e)*d$ssr.df
  L/2 - sum(dnorm(d$ybar, mu, sqrt(sig2a + sig2e/d$n + sig2/(4*d$n) ),
                  log = T) ) }

make.V <- function(th, d) {
# outputs n x dim(th) matrix with rows containing Vs
  sig2a <- exp(th[1])
  sig2e <- exp(th[2])
  sig2 <- exp(th[3])
  mu <- th[4]
  V <- matrix(NA, 12, 4)
  V[1,] <- c(0, 0, d$ss/sig2, 0)
  V[2,] <- c(0, 4, 1, 0) * d$ssr / (sig2 + 4*sig2e)
  S <- sqrt( sig2a + sig2e/d$n + sig2/(4 * d$n) )
  z <- (d$ybar - mu)/S
  V[3:12,-4] <- cbind( z/(2*S), z/(2*S)/d$n, z/(2*S)/(4*d$n) )
  V[3:12,4] <- 1
  V }

phi <- function(th, V, d) {
# outputs vector of length dim(th)
  sig2a <- exp(th[1])
  sig2e <- exp(th[2])
  sig2 <- exp(th[3])
  mu <- th[4]
  S <- sig2a + sig2e/d$n + sig2/(4 * d$n)
  z <- (d$ybar - mu)/S
  L <- - c( 1/sig2, 1/(sig2 + 4*sig2e), 2*z )/2
  apply(V*L, 2, sum) }

volt.init <- c(0, 0, 0, 16)
L <- function(th, d) nlogL( th[1], th[-1], d )
volt.fit <- nlm( L, volt.init, d = volty )
# get initial values for likelihood fit

volt.fr <- fraser.reid( NULL, nlogL, phi, make.V, volt.fit$estimate,
                        d = volty )
volt.ci <- lik.ci( volt.fr )
```

Further topics

Table 7.5 *First and higher order 95% confidence intervals for components of variance.*

	MLE (SE)			Likelihood	Modified likelihood
	First	Higher	Wald pivot	root r	root r^*
$100\sigma^2$	5.41 (0.60)	5.43 (0.61)	(5.35, 6.72)	(4.38, 6.78)	(4.40, 6.81)
$100\sigma_\eta^2$	3.00 (0.83)	3.06 (0.86)	(1.73, 5.20)	(1.66, 5.14)	(1.70, 5.26)
$100\sigma_\alpha^2$	1.02 (0.74)	1.33 (0.95)	(0.24, 4.24)	(0.09, 3.96)	(0.19, 5.39)
$100(\sigma_\alpha^2 + \sigma_\eta^2)$	4.02 (1.04)	4.28 (1.15)	(2.41, 6.68)	(2.41, 7.20)	(2.56, 8.49)
σ_η^2/σ^2	0.55 (0.18)	0.56 (0.18)	(0.30, 1.04)	(0.28, 1.01)	(0.28, 1.03)

expressions in terms of the new parameter of interest ψ and nuisance parameters λ. The R code needed to apply this for estimation of σ_α^2 is shown in Code 7.1. In order to avoid numerical problems with maximization routines, the optimisation uses logarithms of the variance components.

Table 7.5 shows first order and higher order estimates of the parameters, standard errors and 95% confidence limits for the variance parameters and functions of them of possible interest. By analogy with the maximum likelihood estimates, which satisfy the equations $r(\psi) = 0$ for appropriate choices of interest parameters ψ, the higher order estimates are the solutions to $r^*(\psi) = 0$; they are sometimes called median unbiased estimates. The degrees of freedom for σ^2 and for σ_η^2 are quite high, and so the relatively small differences between first and higher order results for them and for their ratio are to be expected. An exact confidence interval for the ratio σ_η^2/σ^2 may be based on the $F_{54,162}$ distribution of

$$\frac{1}{1 + 4\sigma_\eta^2/\sigma^2} \frac{SS^R/54}{SS/162};$$

the resulting 0.95 interval is (0.29, 1.04), very close to the intervals given in Table 7.5. There are just nine degrees of freedom for estimation of σ_α^2, however, leading to much larger higher order corrections. The first order estimate is roughly 25% lower than the higher order one, and first order confidence intervals are appreciably narrower. The confidence intervals were computed on the log scale and then transformed to the original scale; this makes no difference to the intervals based on r and on r^*, which are parametrization-invariant, but avoids difficulties with the intervals based on the Wald pivot, which could include negative values if computed for the variances themselves.

If the population mean were held equal to 16.1 volts, then the voltage of a randomly selected regulator would have mean 16.1 and variance $\sigma_\alpha^2 + \sigma_\eta^2$; thus the percentage lying outside the interval (15.8, 16.4) volts would be $100 \times 2\Phi\{(15.8 - 16.1)/(\sigma_\alpha^2 + \sigma_\eta^2)^{1/2}\}$. Confidence intervals for this can be obtained by transformation of the intervals for $\sigma_\alpha^2 + \sigma_\eta^2$; the higher order interval based on r^* is $(6.1, 30.3)\%$. The corresponding intervals based on t and on r are $(5.3, 23.6)\%$ and $(5.3, 26.3)\%$. There is a substantial correction here largely due to the uncertainty about σ_α^2.

The above analysis is relatively simple owing to the balanced form of the data. For more general discussion see the bibliographic notes.

7.4 Dependent data

Table 7.6 gives the monthly excess returns over the riskless rate for the Acme Cleveland Corporation and for the relevant financial market sector, for the years 1986–1990. The data, shown in the left panel of Figure 7.1, suggest a broadly linear relation between the company return and the market return, though there is a suggestion that the variance increases with the market return.

A standard issue in the analysis of such data is the risk of the security relative to the market, which may be assessed through the value of β in the linear model

$$y_i = \alpha + \beta x_i + \varepsilon_i, \quad i = 1, \ldots, n, \tag{7.4}$$

where x_i is the market excess return, y_i the excess return on the individual security, and the ε_is are error terms, typically supposed to be independent random variables; here $n = 60$. Values of β in excess of $\beta = 1$ indicate that the security is more volatile than the market, and conversely. The line shown with the data in the left panel of Figure 7.1 corresponds to the fitted slope $\widehat{\beta} = 1.13$, with standard error 0.23; any evidence that the company is more volatile than the market is weak. A simple linear regression is typically a poor fit to such data, however. The right panel of Figure 7.1 shows the absolute values of standardized residuals from the fit of (7.4): there is some suggestion that the variance increases with x. Thus the question arises to what extent any variance heterogeneity will affect the estimation of β. A further issue is the possible presence of autocorrelation in the time series of returns: the estimated correlation between successive residuals of -0.28 is significantly different from zero at the 5% level, suggesting that the errors in (7.4) be modelled as a first-order autoregression. Below, we investigate how heteroscedasticity and correlated errors affect inference about β.

To model possible heteroscedasticity, we take the linear model (7.4), but suppose that the error variances are of the form $\mathrm{var}(\varepsilon_i) = \exp(\gamma_0 + \gamma_1 x_i)$; of course equal variances arise when $\gamma_1 = 0$. Such a model could be fitted using the `nlreg` package, but subsequent analysis is a little more direct if higher order inferences are obtained using the discussion following (2.11). In this exponential family model the quantities $z_i = (y_i - \alpha - \beta x_i)e^{-\gamma_0/2 - \gamma_1 x_i/2}$ may be used as pivots, leading to

$$V_i = -(1, x_i, y_i - \widehat{\alpha} - \widehat{\beta}x_i, x_i(y_i - \widehat{\alpha} - \widehat{\beta}x_i)), \quad i = 1, \ldots, n,$$

Table 7.6 *Monthly excess returns (%) over the riskless rate, for the market (Mkt) and for the Acme Cleveland Corporation (Acme), January 1986–December 1990 (Simonoff and Tsai, 1994). The data are here given to one decimal place, but are given to 4 decimals in the data source and in the data frame* return.

Month	1986 Acme	Mkt	1987 Acme	Mkt	1988 Acme	Mkt	1989 Acme	Mkt	1990 Acme	Mkt
January	3.0	−6.1	5.0	7.3	−6.4	1.3	−1.6	−1.1	−12.0	−14.0
February	−16.5	0.8	11.1	−1.2	−11.8	−0.2	−3.8	−9.4	−8.5	−5.9
March	8.0	−0.7	−12.7	−2.7	20.2	−7.3	−7.5	−6.5	−13.0	−5.8
April	−11.0	−6.8	5.5	−4.0	−14.8	−4.3	−10.9	−3.7	−11.7	−10.3
May	−11.5	−0.6	−7.3	−4.8	−17.1	−5.5	−3.7	−4.4	−7.8	2.4
June	−9.9	−4.4	−5.9	−0.2	−1.5	−1.2	2.4	−8.4	−17.0	−7.9
July	−22.7	−11.2	23.6	−0.9	−11.1	−6.2	−7.8	0.3	−7.8	−7.9
August	7.3	3.0	−9.5	−2.1	−16.9	−10.2	−13.2	−5.7	−27.7	−16.1
September	−14.3	−13.0	−13.6	−8.5	−13.6	−3.3	−11.0	−7.9	−20.8	−11.9
October	3.5	0.1	−28.5	−26.2	−8.4	−4.5	−12.6	−10.5	−7.1	−7.6
November	−6.3	−3.4	−17.1	−11	−16.5	−7.9	−9.6	−3.9	−4.6	−0.6
December	−5.9	−7.3	24.3	3.5	15.0	−3.6	6.6	−4.3	−19.1	−2.6

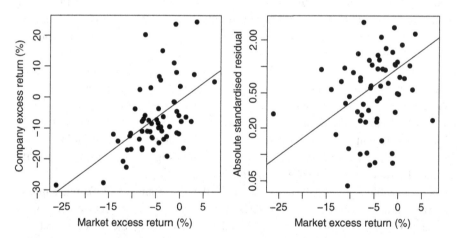

Figure 7.1 Market return data. Left: excess return for the Acme Cleveland Corporation and excess market return, with fitted linear regression. Right: absolute values of standardized residuals and excess market return, with log–linear regression line – note the logarithmic vertical axis.

where $\widehat{\alpha}, \widehat{\beta}$ are the overall maximum likelihood estimates from the original fit. Then we may take

$$\varphi(\theta)^{\mathrm{T}} = -\sum_{i=1}^{n}(y_i - \alpha - \beta x_i)\exp(-\gamma_0 - \gamma_1 x_i) \times V_i,$$

where $\theta^{\mathrm{T}} = (\alpha, \beta, \gamma_0, \gamma_1)$, and apply (2.11) with $\psi = \beta$ and $\lambda^{\mathrm{T}} = (\alpha, \gamma_0, \gamma_1)$. The commands to perform this analysis are given in Code 7.2. The resulting 95% confidence interval for β is shown in Table 7.7, with corresponding intervals from the original fit.

Code 7.2 R code for inference on β based on return data, under heteroscedastic error model.

```
nlogL <- function(psi, lam, d)
{
   sig2 <- exp( lam[1] + lam[2]*d$market )
   mu <- lam[3] + psi*d$market
   - sum( dnorm( d$acme, mu, sqrt(sig2), log = T) )
}

make.V <- function(th, d)
{
   sig2 <- exp( th[2] + th[3]*d$market )
   mu <- th[4] + th[1]*d$market
   z <- d$acme - mu
   cbind( 1, d$market, z, z*d$market )
}

phi <- function(th, V, d)
{
   sig2 <- exp( th[2] + th[3]*d$market )
   mu <- th[4] + th[1]*d$market
   L <- - (d$acme - mu)/sig2
   apply( V*L, 2, sum )
}

return.init <- rep(1, 4)
L <- function(th, d) nlogL( th[1], th[-1], d )
return.fit <- nlm( L, return.init, d=return )
                                    # overall ML fit

return.fr1 <- fraser.reid( NULL, nlogL, phi, make.V,
                    return.fit$estimate, d = return )
return.ci <- lik.ci( return.fr1 )
plot.fr( return.fr1 )
```

Table 7.7 *Excess return data: first order and third order 95% confidence intervals for slope β, based on r and r* respectively, for different error models. The exact interval is based on the Student t_{58} statistic.*

	First order	Third order
Constant variance	(0685, 1.584)	(0.674, 1.591)
Constant variance, exact	(0.676, 1.593)	
Heteroscedastic error	(0.790, 1.441)	(0.757, 1.479)
Autoregressive error	(0.840, 1.633)	(0.812, 1.653)

The same V_i and φ are required to test the hypothesis of variance homogeneity, $\gamma_1 = 0$, for which we set $\psi = \gamma_1$ and $\lambda^{\mathrm{T}} = (\alpha, \beta, \gamma_0)$. The resulting one-tailed P-values based on the Wald pivot t, likelihood root r and modified likelihood root r^* are 0.014, 0.016 and 0.029 respectively, borderline evidence for variance heterogeneity at the two-sided 5% level. Simonoff and Tsai (1994) found weaker evidence, but after removing a possible outlier.

We may model possible serial correlation by supposing that the errors ε_i in (7.4) follow a first-order autoregressive process. As this is a Markov process, the log likelihood may be written as

$$\ell(\theta) = \log f(y_1; \theta) + \sum_{i=2}^{n} \log f(y_i \mid y_{i-1}; \theta),$$

where the first contribution is typically taken from the stationary distribution, and the parameter θ contains the serial correlation ρ in addition to the innovation variance σ^2 and the linear model parameters. Under the assumption of normal errors and apart from additive constants, $\log f(y_i \mid y_{i-1}; \theta)$ equals

$$-\frac{1}{2}\left[\log\{\sigma^2(1-\rho^2)\} + \frac{1}{\sigma^2(1-\rho^2)}\{y_i - \alpha - \beta x_i - \rho(y_{i-1} - \alpha - \beta x_{i-1})\}^2\right],$$

where $\sigma^2 > 0$ and $|\rho| < 1$. As yet there is no comprehensive theory for higher order inference for dependent responses, but one possible approach would be to treat these log likelihood contributions as independent, and to apply (2.11). In this case each of the quantities

$$z_i = \frac{y_i - \alpha - \beta x_i - \rho(y_{i-1} - \alpha - \beta x_{i-1})}{\sigma\sqrt{1-\rho^2}}, \quad i = 2, \ldots, n,$$

is pivotal, conditionally on the preceding response y_{i-1}. Familiar computations yield

$$V_i = -\left(1 - \widehat{\rho}, x_i - \widehat{\rho}x_{i-1}, \widehat{\sigma}\sqrt{1-\widehat{\rho}^2}\,\widehat{z}_i, (y_{i-1} - \alpha - \beta x_{i-1}) + \widehat{\sigma}\,\widehat{\rho}\,\widehat{z}_i/(1-\widehat{\rho}^2)\right),$$

where circumflex indicates evaluation at the overall maximum likelihood estimates, and

$$\varphi(\theta)^{\mathrm{T}} = -\sum_{i=2}^{n} \frac{y_i - \alpha - \beta x_i - \rho(y_{i-1} - \alpha - \beta x_{i-1})}{\sigma^2(1-\rho^2)} \times V_i.$$

The resulting 95% confidence intervals are given in Table 7.7. Throughout we have neglected the initial term of the log likelihood, though in principle it should appear in $\varphi(\theta)$; here there is the difficulty that x_0 is unknown, so the analogue of a stationary distribution of y_1 is unavailable. Note that dropping this log likelihood contribution induces a change of order n^{-1} in the maximum likelihood estimates, so it would seem that the best that is attainable is second order correct inference, even though the responses are continuous.

All the intervals contain $\beta = 1$, but the differences between the first and third order results are larger for both of the enlarged models than for the original model. The third order confidence interval for the original model is very close to the interval found using the exact pivot for $\widehat{\beta}$.

7.5 Vector parameter of interest

Higher order approximation for vector parameters of interest is less well developed than for a scalar parameter of interest. At the time of writing the main methods for improving the χ^2 approximation to the distribution of the likelihood ratio statistic

$$w(\psi) = 2\{\ell(\widehat{\theta}) - \ell(\widehat{\theta}_\psi)\}$$

are Bartlett correction and Skovgaard's w^* approximation; both are described in Section 8.8.

Table 7.8 gives the intervals between failures, in hours, for air-conditioning units on ten different aircraft. These data were originally published by Proschan (1963) and are analysed in detail as Example T of Cox and Snell (1981). We will follow their analysis and illustrate the use of Bartlett correction and Skovgaard's approximation.

We model the failure times by a gamma distribution, allowing separate mean and shape parameters for each aircraft. Writing y_{ij} for the jth observed failure time on aircraft i, where $j = 1, \ldots, n_i$ and $i = 1, \ldots, k$, we have

$$f(y_{ij}; \mu_i, \beta_i) = \frac{1}{\Gamma(\beta_i)} \left(\frac{\beta_i}{\mu_i}\right)^{\beta_i} y_{ij}^{\beta_i - 1} e^{-\beta_i y_{ij}/\mu_i}, \quad y_{ij} > 0, \quad \beta_i, \mu_i > 0,$$

where μ_i and β_i are respectively the mean and shape parameters for failures of the equipment for aircraft i. We shall compare the model with unconstrained β_i and μ_i to that with a common shape parameter, in which $\beta_i = \beta$. The maximum likelihood estimates are given by the equations

$$\widehat{\mu}_i = \bar{y}_{i\cdot}, \quad i = 1, \ldots, k,$$

$$\Psi(\widehat{\beta}_i) - \log\widehat{\beta}_i = \bar{y}_{gi\cdot} - \log\bar{y}_{i\cdot}, \quad i = 1, \ldots, k,$$

$$\Psi(\widehat{\beta}) - \log\widehat{\beta} = \bar{y}_{g\cdot\cdot} - n^{-1}\sum_{i=1}^{k} n_i \log\bar{y}_{i\cdot},$$

Table 7.8 *Failure times (hours) for air-conditioning units in ten aircraft (Cox and Snell, 1981, Example T).*

Aircraft number	Sample size n_i	Failure times y_{ij}
1	23	413 14 58 37 100 65 9 169 447 184 36 201 118 34 31 18 18 67 57 62 7 22 34
2	29	90 10 60 186 61 49 14 24 56 20 79 84 44 59 29 118 25 156 310 76 26 44 23 62 130 208 70 101 208
3	15	74 57 48 29 502 12 70 21 29 386 59 27 153 26 326
4	14	55 320 56 104 220 239 47 246 176 182 33 15 104 35
5	30	23 261 87 7 120 14 62 47 225 71 246 21 42 20 5 12 120 11 3 14 71 11 14 11 16 90 1 16 52 95
6	27	97 51 11 4 141 18 142 68 77 80 1 16 106 206 82 54 31 216 46 111 39 63 18 191 18 163 24
7	24	50 44 102 72 22 39 3 15 197 188 79 88 46 5 5 36 22 139 210 97 30 23 13 14
8	9	359 9 12 270 603 3 104 2 438
9	12	487 18 100 7 98 5 85 91 43 230 3 130
10	16	102 209 14 57 54 32 67 59 134 152 27 14 230 66 61 34

where $\bar{y}_{i\cdot} = \sum_j y_{ij}/n_i$ is the sample mean in the ith group, $\bar{y}_{gi} = \sum_j \log y_{ij}/n_i$ is the logarithm of the geometric mean in the ith group, $n = \sum n_i$ is the total sample size, and $\Psi(\cdot)$ is the digamma function. The maximum likelihood estimates of μ_i and β_i are given in Table 7.9; the maximum likelihood estimate of β assuming a common value is $\widehat{\beta} = 1.006$.

The likelihood ratio statistic

$$w = 2\left[\sum_{i=1}^{k}\left\{-n_i\log\Gamma(\widehat{\beta}_i) + \widehat{\beta}_i\bar{y}_{gi\cdot} + n_i\widehat{\beta}_i\log(\widehat{\beta}_i/\widehat{\mu}_i) - n_i\widehat{\beta}_i\right\} \right.$$
$$\left. - \sum_{i=1}^{k}\left\{-n_i\log\Gamma(\widehat{\beta}) + \widehat{\beta}\bar{y}_{gi\cdot} + n_i\widehat{\beta}\log(\widehat{\beta}/\widehat{\mu}_i) - n_i\widehat{\beta}\right\}\right]$$

has a limiting χ_9^2 distribution as the $n_i \longrightarrow \infty$. The observed value of w is 15.70, leading to a P-value from the χ_9^2 distribution of 0.073; this is weak evidence that the shape parameters are different.

Bartlett adjustment of w entails computing an approximation to the $O(n^{-1})$ term in the expected value of w under the hypothesis of a common shape parameter. The results of this calculation are reported in Cox and Snell (1981) but the details of the computation are omitted. We took an easier, parametric bootstrap, approach of simulating the mean value of w under the common-shape model by simulating failure times y_{ij}^* from Gamma$(\widehat{\beta}, \widehat{\mu}_i)$ densities. The bootstrap mean of w, based on 1 000 samples of the same size as the original data, was 9.78, leading to an estimated Bartlett correction of $9.78/9 = 1.086$ and

Table 7.9 *Maximum likelihood estimates of the gamma parameters. The value* $\widehat{\beta}_3 = 0.91$ *corrects a typographical error in Table T.2 of Cox and Snell (1981); this error is also corrected in Snell (1987).*

i	1	2	3	4	5	6	7	8	9	10
$\widehat{\mu}_i$	95.7	83.5	121.3	130.9	59.6	76.8	64.1	200.0	108.1	82.0
$\widehat{\beta}_i$	0.97	1.67	0.91	1.61	0.81	1.13	1.06	0.46	0.71	1.75

a corrected likelihood ratio statistic $w' = 14.45$, with associated P-value of 0.107, slightly weaker evidence against a common value of β than from the first order analysis.

Skovgaard's correction of w is more directly related to the type of r^* approximation that we use elsewhere in the book. The general expression given at (8.61) is

$$w^* = w\left(1 - \frac{1}{w}\log\gamma\right)^2 ; \tag{7.5}$$

the quantity γ is given at (8.62), but simplifies in a full exponential family model to

$$\gamma = \frac{\{(t - \tilde{\tau})^{\mathrm{T}}\tilde{\Sigma}^{-1}(t - \tilde{\tau})\}^{q/2}}{w^{q/2-1}(\widehat{\theta} - \tilde{\theta})^{\mathrm{T}}(t - \tilde{\tau})}\left(\frac{|\tilde{\Sigma}|}{|\widehat{\Sigma}|}\right)^{1/2}. \tag{7.6}$$

In this expression $t = (\overline{y}_{g1.}, \ldots, \overline{y}_{gk.}, \overline{y}_{1.}, \ldots, \overline{y}_{k.})$ is the vector of sufficient statistics and $\tilde{\tau}$ is the expected value of t evaluated at $\widehat{\beta}, \widehat{\mu}_i$; Σ is the covariance matrix of t, evaluated either at $\widehat{\beta}, \widehat{\mu}_i$ to give $\tilde{\Sigma}$, or at $\widehat{\beta}_i, \widehat{\mu}_i$ to give $\widehat{\Sigma}$; and $q = 9$ is the degrees of freedom in the χ^2 approximation to the distribution of w.

For the data of Table 7.8 we get $\log\gamma = 0.97$, so (7.5) gives $w^* = 13.82$, with an associated P-value of 0.13. This is consistent with the result of using an estimated Bartlett correction.

The code for fitting the gamma model and estimating the mean using the bootstrap is given in Code 7.3; the Skovgaard correction is readily computed from the components of the fit and (7.6).

7.6 Laplace approximation

A different analysis of the failure time data of the previous section uses a hierarchical Bayesian model. For example, we might assume first that the failure times for each aircraft are exponentially distributed, but with a rate or mean parameter that itself has a probability distribution. A hierarchical model would further assume a prior distribution for the parameters of this probability distribution. More specifically, assume

$$f(y_{ij} \mid \lambda_i) = \lambda_i \exp(-\lambda_i y_{ij}), \quad j = 1, \ldots, n_i, \quad i = 1, \ldots, k,$$

Code 7.3 R code for inference on air-conditioning data. The function `fitgamma` fits separate and common gamma models to a data frame of failure times and group number, and uses the `gamma.shape` function in the `MASS` package. The function `bootgamma` gives the parametric bootstrap distribution of the likelihood ratio statistic.

```
fitgamma <- function(data) {
  k <- nlevels( factor(data$number) )
  ydot <- tily <- n <- betahat <- muhat <- rep(0, k)
# Fit separate gamma models
  for(i in 1:k) {
    y <- data$time[data$number==i]
    ydot[i] <- sum(y)
    n[i] <- length(y)
    tily[i] <- sum(log(y))
    muhat[i] <- mean(y)
    betahat[i] <- as.numeric(
                   gamma.shape( glm(y ~ 1, family = Gamma) )[1] )
}
# Fit a common shape parameter
  betall <- as.numeric( gamma.shape(
    glm(data$time ~ factor(data$number) - 1, family = Gamma) )[1] )
# Likelihood ratio statistic
    loglikrat <- 2 * ( sum( - n*lgamma(betahat) +
                           n*betahat*log (betahat/muhat) +
                           betahat*tily - (betahat/muhat)*ydot ) -
                      sum( -n*lgamma(betall) +
                           n*betall*log(betall/muhat) +
                           betall*tily - (betall/muhat)*ydot ) )
    list( k = k, n = n, ydot = ydot, tily = tily, muhat = muhat,
          betahat = betahat, betall = betall, loglikrat = loglikrat )
}

bootgamma <- function(data, fullfit, B) {
  wboot <- rep(0, B)
  num <- data$number
  k <- fullfit$k
  n <- fullfit$n
  betall <- fullfit$betall
  muhat <- fullfit$muhat
  for(b in 1:B) {
  ystar <- as.list(rep(0, k))
    for(i in 1:k) {
       ystar[[i]] <- rgamma( n[i], shape = betall,
                            scale = muhat[i]/betall )
    }
    aircondstar <- data.frame(time = unlist(ystar), number = num)
    wboot[b] <- fitgamma(data = aircondstar)$loglikrat
  }
  wboot
}
```

and further that

$$\pi(\lambda_i \mid \beta, \alpha) = \frac{1}{\Gamma(\alpha)} \beta^\alpha \lambda_i^{\alpha-1} \exp(-\beta\lambda_i),$$

$$\pi(\beta \mid \nu, \phi) = \frac{1}{\Gamma(\nu)} \phi^\nu \beta^{\nu-1} \exp(-\beta\phi).$$

Of course the plausibility of such a model would need careful thought in any application, the key consideration being to what extent there exists an underlying population – here of aircraft – from which random sampling may be supposed to have occurred, so that the parameters for individual aircraft may be regarded as exchangeable.

The posterior marginal density of β is proportional to

$$\beta^{k\alpha+\nu-1} e^{-\phi\beta} \prod_{i=1}^{k} (y_{i\cdot} + \beta)^{-(n_i+\alpha)}, \quad \beta > 0, \tag{7.7}$$

where $y_{i\cdot} = y_{i1} + \cdots + y_{in_i}$, and the normalizing constant is the integral of (7.7) with respect to β. Figure 7.2 shows the Laplace approximation to this density, assuming the fixed values $\alpha = 1.8$, $\nu = 0.1$ and $\phi = 1$, using (8.52), with $\ell(\beta)$ given by the logarithm of (7.7). The maximum of $\ell(\beta)$ is at $\tilde{\beta} = 15.27$, which could serve as an empirical Bayes estimate of β. The r^* approximation to the integral of this density gives $(9.60, 24.43)$ as a posterior probability interval for β; direct numerical integration of the posterior density gives 0.949 as the exact probability content of this interval. We have integrated out the λ_i exactly, so the one-dimensional density to which we apply the Laplace approximation is

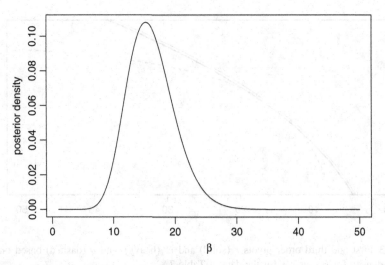

Figure 7.2 Laplace approximation to the posterior density of β given y, for the data of Table 7.8, using expression (8.52).

a marginal density. This is discussed in the context of frequentist inference for a location model at (8.50). A plot of the pivots r, r^* and q based on the marginal density is given in Figure 7.3.

In the frequentist analysis we emphasized the choice between a gamma model with common or different shape parameters for each aircraft. If interest focused instead on the failure rate for one particular aircraft, we could consider the posterior distribution of λ_j, say, marginalizing over the other λs and over β. This is expressed as the ratio of two integrals, both of which can be approximated by a one-dimensional Laplace approximation:

$$\pi_m(\lambda_j \mid y) = \frac{\lambda_j^{n_j+\alpha-1} e^{-y_j \cdot \lambda_j}}{\Gamma(n_j+\alpha)} \frac{\int_0^\infty \exp\{\ell_j(\beta)\}d\beta}{\int_0^\infty \exp\{\ell(\beta)\}d\beta},$$

where

$$\ell_j(\beta) = (k\alpha+\nu-1)\log(\beta) - \beta(\lambda_j+\phi) - \sum_{i\neq j}(n_i+\alpha)\log(y_{i\cdot}+\beta)$$

and $\ell(\beta)$ is the logarithm of (7.7).

The numerator of π_m must be approximated for a range of values for λ_j; the posterior mean of this approximation, for example, gives an empirical Bayes estimate for λ_j. For λ_8, the aircraft with the smallest number of observations and smallest direct estimate of the rate parameter, this empirical Bayes estimate is 0.006 compared to the direct estimate of 0.005.

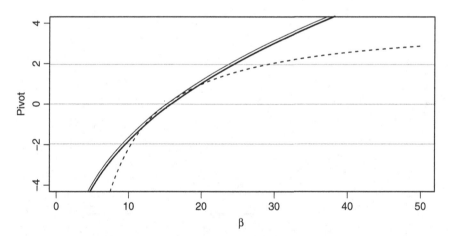

Figure 7.3 First and third order pivots r (solid) and r^* (heavy), and q (dashed) based on the log marginal density for β given y, for the data of Table 7.8.

7.7 Partial likelihood

One of the most widely used models for survival data is the proportional hazards model. In its simplest form the failure times $T_1, \ldots, T_n$ for n independent individuals have survivor functions

$$\Pr(T_i > t) = \exp\{-\xi_i H(t)\}, \quad i = 1, \ldots, n,$$

where ξ_i is a non-negative function of parameters β, and $H(t)$, a so-called baseline cumulative hazard function, is a non-negative increasing function of t. In applications it is usual to take $\xi_i = \exp(x_i^T \beta)$, where x_i is a vector of covariates for the ith individual and β is a vector of unknown parameters. If the derivative $h(t) = \mathrm{d}H(t)/\mathrm{d}t$ exists, then the probability density corresponding to failure of the ith individual at time t is $\exp\{-\xi_i H(t)\} \xi_i h(t)$.

Now suppose that the failure times are subject to right-censoring by a mechanism independent of their values and uninformative about their distribution. Let d_i be an indicator of the survival status of the ith individual, with $d_i = 1$ if the observed value t_i of T_i is a failure time and $d_i = 0$ if t_i represents a right-censored value, that is, if $T_i > t_i$. Thus the data consist of triples $(x_1, t_1, d_1), \ldots, (x_n, t_n, d_n)$, and the likelihood for β may be written as

$$\prod_{i=1}^n \exp\{-\xi_i H(t_i)\} \{\xi_i h(t_i)\}^{d_i}.$$

It might appear that information about β cannot be extracted without further assumptions about the form of H, but an ingenious argument leads to the so-called partial likelihood

$$L(\beta) = \prod_i \frac{\xi_i}{\sum_{j \in \mathcal{R}_i} \xi_j}, \tag{7.8}$$

where the product is taken over those individuals that are observed to fail, and $\mathcal{R}_i$ represents the risk set comprising those individuals still available to fail at time t_i. In fact $L(\beta)$ can be regarded as a marginal likelihood, based on the ranks of the observed failure times. When $\xi_i = \exp(x_i^T \beta)$, the individual components of (7.8) have the form

$$\frac{\exp(x_i^T \beta)}{\sum_{j \in \mathcal{R}_i} \exp(x_j^T \beta)},$$

and $L(\beta)$ strongly resembles an exponential family likelihood, though it is not an exponential family because of the successive conditioning implicit in (7.8) and the censoring. Despite this one might hope that the higher order procedures described in Section 8.6.1 would yield useful improvements to first order likelihood procedures. The numerical experiment described below suggests that this is not the case.

In order to test the usefulness of higher order asymptotics for the proportional hazards model, we conducted two simulation experiments. In the first we generated two groups

Table 7.10 *Normality of likelihood root r, modified likelihood root r*, and modified Wald pivot t, in fits of proportional hazards model to 10 000 simulated datasets of size n = 20. The upper figures are for a model without nuisance parameters, and the lower ones are for a model with four nuisance parameters. SE is the simulation standard error.*

p	1	2.5	5	10	20	50	80	90	95	97.5	99
$\Phi(r)$	1.2	3.5	6.5	12.4	23.4	52.6	80.0	89.5	94.7	97.3	98.9
$\Phi(r^*)$	1.3	3.7	6.8	13.1	24.3	54.0	81.0	90.2	95.1	97.7	99.0
$\Phi(t)$	1.6	4.3	7.3	13.5	24.0	52.6	80.8	90.8	96.2	98.5	99.7
$\Phi(r)$	3.2	5.7	9.2	15.0	25.2	51.1	75.9	85.4	91.0	94.6	97.2
$\Phi(r^*)$	2.7	5.1	8.4	14.2	24.6	51.4	76.8	86.5	92.0	95.3	97.8
$\Phi(t)$	0.4	2.4	6.4	13.6	24.7	51.1	76.7	88.0	94.7	98.5	99.8
SE	0.1	0.2	0.2	0.3	0.4	0.5	0.4	0.3	0.2	0.2	0.1

of standard exponential variables, one of size $n_1 = 5$ and the other of size $n_2 = 15$. Five observations in each group were left uncensored, and the remaining ten in the second group were subjected to random censoring by variables generated from the uniform distribution on the interval $(0, 4)$, about 10% censoring overall. The proportional hazards model was fitted using the partial likelihood (7.8) to estimate the scalar parameter β which corresponds to membership of the second group; the true value taken was $\beta = 0$. Such datasets were generated 10 000 times, and the corresponding likelihood root r, the Wald pivot $t = (\widehat{\beta} - \beta) j_p(\widehat{\beta})^{1/2}$ and the modified likelihood root $r^* = r + \log(t/r)/r$ were computed in each case. The upper part of Table 7.10 shows the empirical proportions of simulated values $\Phi(r)$, $\Phi(r^*)$ and $\Phi(t)$, less than or equal to various values of p. For a quantity with the standard normal distribution, this proportion would equal p, apart from simulation variability. There is no clear advantage to the use of r^* rather than r, though both seem to have somewhat better behaviour than t, which is not parametrization-invariant.

The lower part of the table shows similar results for samples of size $n = 20$ with five covariates generated using standard normal variables and a similar censoring pattern to that used above, but about 30% censoring. In this case the nuisance parameter adjustment (8.46) must be included; its effect is to adjust the Wald statistic, whose performance is very similar to that of r, in the correct direction, so that t gives generally better inferences than either r or r^*. Figure 7.4 shows normal probability plots of t and of r; that for r^* is very close to that for r. The distribution of t is rather non-normal, but near ± 2 its quantiles are close to those of the standard normal distribution, accounting for the results in Table 7.10. The distribution of r is close to normal, but as its variance exceeds one, confidence intervals based on r will tend to be too short.

Data sets in survival analysis are rarely as small as those considered here, so it seems likely that higher order inference for the proportional hazards model will only rarely be worth considering.

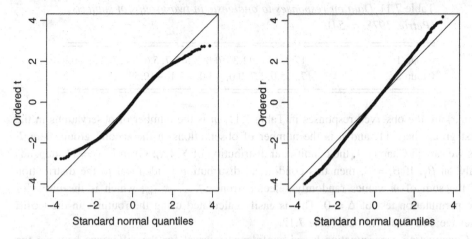

Figure 7.4 Normal probability plots of simulated values of t and r based on proportional hazards model. The diagonal lines indicate a perfect fit of the standard normal distribution.

7.8 Constructed exponential families

Higher order approximation can be used for inference in a number of problems using constructed exponential families and saddlepoint approximations for these families. For example, the bootstrap distribution of a sample mean is obtained from the multinomial distribution of the resampling indices. The multinomial distribution in turn can be obtained as the conditional distribution in independent Poisson sampling, conditional on the total number of events. The Poisson model is the constructed exponential family, and the higher order likelihood approximation based on r^* is identical to the saddlepoint approximation to the bootstrap distribution of the mean.

In this section we will illustrate the use of this construction to approximate the permutation distribution for a two-sample comparison, following Booth and Butler (1990). Table 7.11 shows the responses for two groups of subjects given an analgesic: the numbers reported are the differences in pain response for each subject, before and after the analgesic. Of interest is whether the reduction is similar for the two groups of subjects.

We model the responses in group 2 as the realization of a sample of independent, identically distributed variables from a distribution $F(\cdot)$, say, and those in group 1 as a similar sample from the distribution $F(\cdot + \Delta)$. One way to test the hypothesis of no difference between the two groups non-parametrically is to use the permutation test for $\Delta = 0$. This treats the observed responses as fixed and considers all possible assignments of these responses to two groups of 11 and 7 subjects. We can do this indirectly by constructing a Bernoulli response vector $(y_1, \ldots, y_{m+n})$ to index the two groups, where $\Pr(Y_i = 1 \mid x_i) = \theta_i$, say, and

$$\log\left(\frac{\theta_i}{1-\theta_i}\right) = \beta_0 + \beta_1 x_i, \quad i = 1, \ldots, m+n.$$

Table 7.11 *Data on responses to analgesic in two groups of subjects (Petrie, 1978, p. 54).*

Group 1	17.9, 13.3, 11.3, 10.6, 7.6, 5.7, 5.6, 5.4, 3.3, 3.1, 0.9
Group 2	7.7, 5.0, 1.7, 0.0, −3.0, −3.1, −10.5

The x_i are the observed responses in Table 7.11, m is the number of observations in the first group, here 11, and n is the number of observations in the second group, here 7. As we saw in Chapter 4, the conditional distribution of $\sum y_i x_i$, given $\sum y_i = 11$, depends only on β_1. If $\beta_1 = 0$, then this conditional distribution is identical to the distribution of the sum of m values randomly selected from $x_1, \ldots, x_{n+m}$, which is the basis for the permutation test of $\Delta = 0$. This is easily calculated using the routines in the `cond` package; see Code 7.4 and Table 7.12.

Constructing a permutation-based confidence interval for the difference between the two groups involves using the `cond` function iteratively. Upper and lower confidence limits at level $\alpha/2$ are obtained by finding values Δ_L and Δ_U, say, for which two permutation tests, based on $x_1 - \Delta_L, \ldots, x_m - \Delta_L, x_{m+1}, \ldots, x_{m+n}$, and on $x_1 - \Delta_U, \ldots, x_m - \Delta_U, x_{m+1}, \ldots, x_{m+n}$, have P-value exactly equal to $\alpha/2$. As the permutation distribution of the sample mean is in large samples equivalent to the two-sample t-test, this provides useful starting values for the iteration to determine Δ_L and Δ_U.

Code 7.4 Extraction of components of the `summary.cond` object for computing confidence intervals for the permutation distribution.

```
> boothb  # data frame with response and group indicator
    Group y
1  17.9  1
2  13.3  1
3  11.3  1
...
18 -10.5 0

> booth.glm <- glm( y ~ Group, family = binomial, data = boothb )
> booth.cond <- cond.glm( booth.glm, offset = Group )
> booth.summary <- summary( booth.cond, test = 0 )
# attributes( booth.summary ) lists the components of this object
> booth.summary$signif.tests$stats[1:4, 2]
Wald pivot
                        0.024167532
Wald pivot (cond. MLE)
                        0.026817569
Likelihood root
                        0.001661714
Modified likelihood root
                        0.003352744
```

```
# Now we iterate "cond" over plausible range of lower limits

> delta.low <- seq(1.5, 2, length = 30)
> conf.low <- matrix(0, nrow = length(delta.low), ncol = 3)

> for( i in 1:length(delta.low) ) {
+    shift <- boothb$Group - delta.low[i]*boothb$y
+    shift.cond <- cond( glm( boothb$y ~ shift, family = binomial ),
+                           offset = shift )
+    shift.cond.summary <- summary( shift.cond,
+                                    test=0 )$signif.tests$stats
+    conf.low[i,1] <- as.numeric( shift.cond.summary[4,2] )
+    conf.low[i,2] <- as.numeric( shift.cond.summary[3,2] )
+    conf.low[i,3] <- as.numeric( shift.cond.summary[1,2] ) }

> predict( smooth.spline(conf.low[,1], delta.low), 0.025 )$y
[1] 2.558247
> predict( smooth.spline(conf.low[,2], delta.low), 0.025 )$y
[1] 3.379706
> predict( smooth.spline(conf.low[,3], delta.low), 0.025 )$y
[1] 0.6215863
```

Table 7.12 *Permutation test P-values and confidence intervals for the mean (or median[+]) difference between groups 1 and 2, for data in Table 7.11.*

Method	P-value	95% confidence interval
Standard t-test	0.0133	(2.02, 14.01)
[+]Wilcoxon rank sum test	0.0114	(1.60, 13.80)
Permutation t-test	0.0073	(2.37, 13.64)
Constructed exponential family, r^*	0.0067	(2.56, 13.83)
Constructed exponential family, r	0.0033	(3.38, 15.03)
Constructed exponential family, Wald pivot	0.0483	(0.62, 10.08)

Bibliographic notes

The use of higher order likelihood inference in calibration problems is treated in detail in Bellio (2000, Chapter 3); see also Bellio (2003). Eno (1999) discusses Bayesian calibration with reference and other non-informative priors. Bellio (2000) also treats multivariate calibration, and calibration in a random effects model.

Christensen *et al.* (2006) build on Jensen (1997) to discuss higher order inference for variance component models. In general it is necessary to re-interpret the variance components as covariances, in order to allow the possibility of negative estimates, and to take care with the log likelihood maximization when the parameter of interest is a linear combination of variance components, as would often be the case. In unbalanced cases there is no low-dimensional minimal sufficient statistic and a different argument is needed to produce a marginal likelihood on which to base inference. Christensen *et al.* (2006) give expressions for the quantities required for the general balanced case and for the unbalanced one-way layout, and simulations showing the excellent performance of the approach; see also Lyons and Peters (2000). Davison (2003, Section 9.4) contains a brief account of variance components, and Cox and Solomon (2003) give a fuller treatment. Median unbiased estimation is discussed by Pace and Salvan (1999).

There is some limited literature on the use of higher order asymptotics in mixed linear and nonlinear models; see the references in Problems 46 and 47.

Simonoff and Tsai (1994) discuss how adjusted profile likelihoods may be used to provide improved tests for heteroscedasticity in linear regression. Score tests for this purpose were proposed by Cook and Weisberg (1983) and are particularly useful in econometric applications. As is often the case with tests on variances, it is here important to make the equivalent of a degrees of freedom adjustment; indeed Bartlett (1937) introduced his correction in a similar context. See also Box (1949) and papers by F. Cribari-Neto, S. L. P. Ferrari and colleagues (e.g. Cribari-Neto and Ferrari, 1995; Cribari-Neto *et al.*, 2000; Ferrari *et al.*, 2001) on this and corrections for sandwich covariance matrices in the presence of heteroscedasticity.

Explicit expressions for the Bartlett correction for inference about the mean or the shape parameter of a gamma distribution are given in Cordeiro and Aubin (2003), for the single sample problem; see also Ferrari *et al.* (2005), where expressions for the Bartlett correction based on adjusted likelihoods are also considered. If the parameter of interest is scalar, then Bartlett adjustment is equivalent to standardizing the likelihood root r by estimating or approximating its mean and variance; see Problem 16. The Bayesian analysis of Section 7.6 follows Davison (2003, Chapter 11), who also compares Laplace approximation to Gibbs sampling.

The numbers in Table 7.12 differ slightly from those reported by Lehmann (1975, p. 92) and Booth and Butler (1990), as our analysis uses all 18 observations from the data source (Petrie, 1978). Lehmann deleted one observation at random to reduce the largest group size to ten.

The saddlepoint approximation to the bootstrap distribution of the sample mean, and of regular estimating equations, was first discussed in Davison and Hinkley (1988). Booth and Butler (1990) expand on this discussion, and in particular consider both the so-called *double saddlepoint*, which is essentially the exponential family version based on the profile log likelihood, with the *sequential saddlepoint*, the approximation based on the adjusted profile log likelihood; see also Section 8.6.1. Booth and Butler (1990) also show how the Bayesian bootstrap can be developed by constructing a gamma regression model.

Butler *et al.* (1992a, 1992b) illustrate the use of the embedding exponential family technique to derive approximations to a number of test statistics arising in multivariate analysis. See also Butler (2007). One example is outlined in more detail in Problem 48.

The use of higher order methods in censored data is not straightforward, with the exception of Type II censored data discussed in Sections 5.5 and 5.6. DiCiccio and

Martin (1993) illustrate the use of a version of r^* motivated by Bayesian arguments for inference about the mean of an exponential distribution with Type I censoring. There is a vast literature on the proportional hazards model: Davison (2003, Chapter 10) contains a brief account, with fuller treatments given by Cox and Oakes (1984), Fleming and Harrington (1991), Andersen *et al.* (1993), Klein and Moeschberger (1997), and Therneau and Grambsch (2000).

8

Likelihood approximations

8.1 Introduction

In this chapter we give a brief overview of the main theoretical results and approximations used in this book. These approximations are derived from the theory of higher order likelihood asymptotics. We present these fairly concisely, with few details on the derivations. There is a very large literature on theoretical aspects of higher order asymptotics, and the bibliographic notes give guidelines to the references we have found most helpful.

The building blocks for the likelihood approximations are some basic approximation techniques: Edgeworth and saddlepoint approximations to the density and distribution of the sample mean, Laplace approximation to integrals, and some approximations related to the chi-squared distribution. These techniques are summarized in Appendix A, and the reader wishing to have a feeling for the mathematics of the approximations in this chapter may find it helpful to read that first.

We provide background and notation for likelihood, exponential family models and transformation models in Section 8.2 and describe the limiting distributions of the main likelihood statistics in Section 8.3. Approximations to densities, including the very important p^* approximation, are described in Section 8.4. Tail area approximations for inference about a scalar parameter are developed in Sections 8.5 and 8.6. These tail area approximations are illustrated in most of the examples in the earlier chapters. Approximations for Bayesian posterior distribution and density functions are described in Section 8.7. Inference for vector parameters, using adjustments to the likelihood ratio statistic, is described in Section 8.8.

8.2 Likelihood and model classes

We assume we have a parametric model for a random vector $Y = (Y_1, \ldots, Y_n)$, which has a probability density or mass function $f(y; \theta)$, with $\theta \in \mathbb{R}^d$. The likelihood function is

$$L(\theta) = L(\theta; y) = c(y)f(y; \theta) \propto f(y; \theta). \tag{8.1}$$

This is expressed as a function of θ, for fixed y, and thus the constant of proportionality in (8.1) can be a function of y, but not of θ. In considering the distributional properties of the likelihood function, we implicitly or explicitly use the notation $L(\theta; Y)$. If the components of Y are independent, then the likelihood is a product of their density functions, so it is often more natural to work with the log likelihood function

$$\ell(\theta) = \ell(\theta; y) = \log L(\theta; y) = \log f(y; \theta) + a(y).$$

Throughout the book we ignore additive constants appearing in log likelihood functions.

The *maximum likelihood estimate* of θ, denoted by $\widehat{\theta}$, if it exists, is the value of θ that maximizes $L(\theta)$ or equivalently $\ell(\theta)$. In many models it is found by solving the *score equation*

$$\ell_\theta(\widehat{\theta}; y) = 0, \tag{8.2}$$

where $\ell_\theta(\theta; y) = \partial\ell(\theta; y)/\partial\theta$ is the score function. By convention θ is a column vector; thus so is ℓ_θ. Both the maximum likelihood estimate and the score function are also functions of y, and their sampling distributions are determined by the distribution of Y. When the components of Y are independent the score function is a sum of independent terms, so we might expect a central limit theorem to hold under some conditions; the limiting normal distribution is characterized by its mean and covariance matrix. The $d \times d$ covariance matrix of $\ell_\theta(\theta)$ is called the *expected Fisher information*

$$i(\theta) = \text{cov}\{\ell_\theta(\theta)\}.$$

The *observed Fisher information function* is the $d \times d$ matrix

$$j(\theta) = -\ell_{\theta\theta}(\theta) = -\frac{\partial^2\ell(\theta)}{\partial\theta\partial\theta^{\mathrm{T}}},$$

and the observed Fisher information is $j(\widehat{\theta})$.

We are deliberately vague about regularity conditions; references are given in the bibliographic notes. In general we assume that the log density of Y is smooth and unimodal in θ, so that the score equation defines the maximum likelihood estimate uniquely, that the support of the density does not depend on any component of θ, and that we can freely exchange integration and differentiation with respect to θ. Then we can use the identity

$$\int f(y; \theta)\mathrm{d}y = 1$$

to verify the *Bartlett identities*

$$\mathrm{E}\{\ell_\theta(\theta)\} = 0, \qquad \mathrm{E}\{\ell_{\theta\theta}(\theta)\} + \text{cov}\{\ell_\theta(\theta)\} = 0,$$

and thus

$$\mathrm{E}\{j(\theta)\} = i(\theta). \tag{8.3}$$

In view of (8.3), some books define the expected Fisher information to be $\mathrm{E}\{-\ell_{\theta\theta}(\theta)\}$. The Bartlett identities can be extended to higher order derivatives.

We often consider inference for a component of θ, with the remaining components treated as nuisance parameters. We write $\theta = (\psi, \lambda)$, where ψ is the parameter of interest. In some problems the nuisance parameter might be available only implicitly, but it is typically useful at least conceptually to assume that it is explicitly available. In reparametrizing the model, from for example (θ_1, θ_2) to (ψ, λ), it is important to ensure that the parameter space for θ, often denoted by Θ, is preserved by the transformation. In particular, when $\Theta = \Psi \times \Lambda$, where Ψ and Λ are the corresponding parameter spaces for the parameter of interest and the nuisance parameter, we say that the (ψ, λ) parametrization is variation-independent. If the allowable values of λ were to depend on ψ, this would introduce irregularities into the model and the usual asymptotic theory would not apply.

The partitioning of the parameter entails a similar partitioning of the score vector

$$\ell_\theta(\theta) = \begin{pmatrix} \ell_\psi(\theta) \\ \ell_\lambda(\theta) \end{pmatrix},$$

the Fisher information matrix

$$i(\theta) = \begin{pmatrix} i_{\psi\psi}(\theta) & i_{\psi\lambda}(\theta) \\ i_{\lambda\psi}(\theta) & i_{\lambda\lambda}(\theta) \end{pmatrix},$$

and the observed Fisher information. The inverse matrices are similarly partitioned, an example being

$$i(\theta)^{-1} = \begin{pmatrix} i^{\psi\psi}(\theta) & i^{\psi\lambda}(\theta) \\ i^{\lambda\psi}(\theta) & i^{\lambda\lambda}(\theta) \end{pmatrix}.$$

The *constrained maximum likelihood estimate* of λ with ψ fixed is denoted $\widehat{\lambda}_\psi$, and in regular problems is determined by

$$\ell_\lambda(\psi, \widehat{\lambda}_\psi) = 0.$$

The *profile log likelihood* is

$$\ell_\mathrm{p}(\psi) = \ell(\psi, \widehat{\lambda}_\psi)$$

with associated observed profile information function

$$j_\mathrm{p}(\psi) = -\partial^2 \ell_\mathrm{p}(\psi)/\partial\psi\partial\psi^\mathrm{T}.$$

Standard results for partitioned matrices give

$$|j_\mathrm{p}(\psi)| = |j(\psi, \widehat{\lambda}_\psi)|/|j_{\lambda\lambda}(\psi, \widehat{\lambda}_\psi)| \tag{8.4}$$

and

$$j_\mathrm{p}(\psi) = \{j^{\psi\psi}(\psi, \widehat{\lambda}_\psi)\}^{-1}.$$

We often use the shorthand notation $\widehat{\theta}_\psi = (\psi, \widehat{\lambda}_\psi)$.

In addition to derivatives on the parameter space, we also have occasion to use derivatives of the log likelihood on the sample space. For example, to define the canonical parameter φ of the tangent exponential model as at (2.12), we use

$$\ell_{;y}(\theta; y^0) = \left.\frac{\partial\ell(\theta; y)}{\partial y}\right|_{y=y^0}.$$

In models where we can express y explicitly as a one-to-one function of $(\widehat{\theta}, a)$, we use notation such as

$$\ell_{;\widehat{\theta}}(\theta; \widehat{\theta}, a) = \frac{\partial\ell(\theta; \widehat{\theta}, a)}{\partial\widehat{\theta}}, \quad \ell_{\theta;\widehat{\theta}}(\theta; \widehat{\theta}, a) = \frac{\partial^2\ell(\theta; \widehat{\theta}, a)}{\partial\theta\partial\widehat{\theta}^\mathsf{T}}.$$

The functions $\ell_{;\widehat{\theta}}$ and $\ell_{;y}$ are *sample space derivatives*, while $\ell_{\theta;\widehat{\theta}}$ is a mixed derivative in both parameter and sample spaces.

The family of distributions $\{\mathcal{F}_\theta; \theta \in \Theta \subset R^d\}$ for a single random variable Y is called an exponential family model if the distributions have densities of the form

$$f(y; \theta) = \exp\left\{\sum_{i=1}^{d}\varphi_i(\theta)v_i(y) - c(\theta) - h(y)\right\}, \quad \theta \in \Theta. \tag{8.5}$$

The parameter φ is called the canonical parameter of the model, and $v(y)$ is the sufficient statistic. The canonical form of the model is

$$f(y; \varphi) = \exp\{\varphi^\mathsf{T}v(y) - c(\varphi) - h(y)\},$$

where $c(\varphi) = c\{\varphi(\theta)\}$. The exponential family is said to have full rank if φ can take any value for which $\int\exp\{\varphi^\mathsf{T}v(y) - h(y)\}dy < \infty$.

Exponential families are closed under sampling: if $(y_1, \ldots, y_n)$ are independent observations from model (8.5) then $\sum v(y_i)$ is the sufficient statistic based on the sample, and its distribution is also an exponential family model. Exponential families are also closed under conditioning, as described in Section 8.6.1.

A *curved exponential family* is obtained when the components of φ satisfy one or more nonlinear constraints. In that case φ belongs to a manifold of lower dimension than that of the sufficient statistic v.

Exponential family models form the basis for generalized linear models, and are used extensively throughout the book, particularly in Chapter 4. The general expression for q given at (2.11) is based on the canonical parameter φ of an approximating exponential family model, as described in more detail in Section 8.4.2.

Transformation family models are defined, as the name suggests, through a group of transformations on the sample space. Denote by g a given transformation of the response y, and suppose that $f(gy; \theta) = f(y; g^*\theta)$ for some transformation g^* of the parameter. If the set of possible values for θ and $g^*\theta$ are identical, then the family of densities $\{f(y; \theta), \theta \in R^d\}$

is invariant under the transformation g. This family is called a transformation family when this property holds for all g in a group of transformations $\mathcal{G}$. It can be shown that $\mathcal{G}^* = \{g^* \text{ induced by } g \in \mathcal{G}\}$ is a group in the parameter space.

Most of the models discussed in Chapter 5 are transformation models for a suitably defined group of transformations: particular examples are location, scale, location–scale and regression–scale models.

8.3 First order theory

8.3.1 Scalar θ

This section expands on the brief discussion in Section 2.2. We assume that $y = (y_1, \ldots, y_n)$ is a sample of independent, identically distributed random variables. Then

$$\ell(\theta; y) = \sum \log f(y_i; \theta),$$

where by a slight abuse of notation we have also used $f(\cdot; \theta)$ for the density of a single observation. Also $i(\theta) = n i_1(\theta)$, say, where $i_1(\theta)$ is the expected Fisher information in a single observation. Let $\overset{\mathrm{d}}{\to}$ denote convergence in distribution. Then, under some regularity conditions on f, we have a central limit theorem for the score function, that is

$$i(\theta)^{-1/2} \ell_\theta(\theta) \overset{\mathrm{d}}{\to} N(0, 1),$$

and this forms the basis for deriving the asymptotic distribution of the maximum likelihood estimator and the likelihood ratio statistic. For by expansion of the score equation (8.2) in a Taylor series about θ we can obtain

$$(\widehat{\theta} - \theta) i(\theta)^{1/2} = i(\theta)^{-1/2} \ell_\theta(\theta) \{1 + o_p(1)\},$$

and a similar expansion of the log likelihood function yields

$$w(\theta) = 2\{\ell(\widehat{\theta}) - \ell(\theta)\} = (\widehat{\theta} - \theta)^2 i(\theta) \{1 + o_p(1)\}.$$

The notation $o_p(1)$ indicates a random variable that converges in probability to 0; see Appendix A. These expressions also assume that $\widehat{\theta}$ is consistent for θ, and use the result that $n^{-1}\{j(\widehat{\theta}) - i(\theta)\}$ converges in probability to 0, by the weak law of large numbers.

From these limiting results we have, as at (2.1), (2.2) and (2.3), the following three first order approximations:

$$r(\theta) = \mathrm{sign}(\widehat{\theta} - \theta) \sqrt{[2\{\ell(\widehat{\theta}) - \ell(\theta)\}]} \overset{\cdot}{\sim} N(0, 1), \tag{8.6}$$

$$s(\theta) = j(\widehat{\theta})^{-1/2} \ell_\theta(\theta) \overset{\cdot}{\sim} N(0, 1), \tag{8.7}$$

$$t(\theta) = j(\widehat{\theta})^{1/2} (\widehat{\theta} - \theta) \overset{\cdot}{\sim} N(0, 1). \tag{8.8}$$

The three quantities r, s and t figure prominently in the theory outlined in the following sections and in the higher order approximations of earlier chapters. The left-hand side of (8.7) is the standardized score function, also called the score pivot. That of (8.8) is the standardized maximum likelihood estimate, also called the Wald pivot. Finally we call r the *likelihood root*, though other authors have termed it the *signed likelihood ratio statistic* or the *directed likelihood ratio statistic*. To the first order of asymptotic theory these three quantities all have the same distribution, although they usually have different numerical values in any particular application. Numerical examples are given in Section 2.1 and Section 3.2.

8.3.2 Vector θ

If θ is a vector of fixed length $d < n$, similar asymptotic results apply with only notational changes. It is easier to describe the limiting results as $n \to \infty$ using the quadratic forms of the relevant statistics, that is,

$$w(\theta) \xrightarrow{d} \chi_d^2,$$

$$\ell_\theta^{\mathrm{T}}(\theta) i(\theta)^{-1} \ell_\theta(\theta) \xrightarrow{d} \chi_d^2,$$

$$(\widehat{\theta} - \theta)^{\mathrm{T}} i(\theta)(\widehat{\theta} - \theta) \xrightarrow{d} \chi_d^2.$$

When $\theta = (\psi, \lambda)$ is partitioned into a parameter of interest ψ of length q, and a nuisance parameter λ of length $d - q$, we have

$$w(\psi) = 2\{\ell(\widehat{\psi}, \widehat{\lambda}) - \ell(\psi, \widehat{\lambda}_\psi)\} \xrightarrow{d} \chi_q^2, \tag{8.9}$$

$$\ell_\psi^{\mathrm{T}}(\psi, \widehat{\lambda}_\psi) j^{\psi\psi}(\widehat{\psi}, \widehat{\lambda}) \ell_\psi(\psi, \widehat{\lambda}_\psi) \xrightarrow{d} \chi_q^2, \tag{8.10}$$

$$(\widehat{\psi} - \psi)^{\mathrm{T}} \{j^{\psi\psi}(\widehat{\psi}, \widehat{\lambda})\}^{-1} (\widehat{\psi} - \psi) \xrightarrow{d} \chi_q^2. \tag{8.11}$$

In the special case that $q = 1$, we also have

$$r(\psi) = \mathrm{sign}(\widehat{\psi} - \psi) \sqrt{[2\{\ell_{\mathrm{p}}(\widehat{\psi}) - \ell_{\mathrm{p}}(\psi)\}]} \ \dot{\sim} \ N(0, 1),$$

$$s(\psi) = j_{\mathrm{p}}(\widehat{\psi})^{-1/2} \ell_\psi(\psi, \widehat{\lambda}_\psi) \ \dot{\sim} \ N(0, 1),$$

$$t(\psi) = j_{\mathrm{p}}(\widehat{\psi})^{1/2} (\widehat{\psi} - \psi) \ \dot{\sim} \ N(0, 1).$$

In (8.7) and (8.8), and in the nuisance parameter versions above, we have standardized by the observed Fisher information, although a version of these statistics standardized instead by the expected Fisher information, $i(\theta)$, or indeed by anything that consistently estimates $i(\theta)$, such as $i(\widehat{\theta})$, could be used without changing the limiting properties of the statistic. In particular, the score statistic (8.10) is often standardized by $j^{\psi\psi}(\psi, \widehat{\lambda}_\psi)$, which avoids calculation of the overall maximum likelihood estimate. However, the theory of higher order asymptotics leads us to recommend standardization by the observed Fisher information at $\tilde{\theta}$ for first order approximations.

If y is a vector of independent but not identically distributed components, as often arises in a regression setting, then to obtain results analogous to those above we need a central limit theorem for the score function, which in turn will require that $i(\theta) \to \infty$ as $n \to \infty$. If the components of y are dependent, the situation is more difficult, and needs to be treated on a case by case basis. For example, in the analysis of grouped or doubly-indexed data y_{ij}, we usually need the number of observations per group to increase at a faster rate than the number of groups.

8.4 Higher order density approximations

8.4.1 The p^* approximation

The limiting results of the previous subsection lead to a d-dimensional normal approximation to the density of $\widehat{\theta}$, with mean θ and covariance matrix $j(\widehat{\theta})^{-1}$. This is a first order approximation, in the sense that it has relative error $O(n^{-1/2})$.

The basic higher order approximation to the density of the maximum likelihood estimator is the so-called p^* *approximation*:

$$f(\widehat{\theta} \mid a; \theta) \doteq c|j(\widehat{\theta})|^{1/2} \exp\{\ell(\theta) - \ell(\widehat{\theta})\} \tag{8.12}$$
$$\equiv p^*(\widehat{\theta} \mid a; \theta).$$

In the right-hand side of (8.12) the dependence of the log likelihood function and the observed Fisher information on the data has been suppressed, but it is very important for the interpretation of the p^* approximation. In fuller notation

$$p^*(\widehat{\theta} \mid a; \theta) = c(\theta, a)|j(\widehat{\theta}; \widehat{\theta}, a)|^{1/2} \exp\{\ell(\theta; \widehat{\theta}, a) - \ell(\widehat{\theta}; \widehat{\theta}, a)\}, \tag{8.13}$$

where we have expressed the data vector as $y = (\widehat{\theta}, a)$, a transformation that we assume is one-to-one. The factor $c(\theta, a)$ is a normalizing constant. The transformation of y is needed for the correct interpretation of (8.12), as the left hand side is a d-dimensional density, whereas the sample space is n-dimensional. We have reduced the dimension of the sample space to the dimension of the parameter, d, by conditioning on the statistic a. In order that this be both a valid approximation and useful for inference, it is necessary that this conditioning not lose information about the parameter θ. One way to ensure this is to require a to have a distribution free of θ, or in other words to be an *ancillary statistic*. Although an exactly ancillary statistic does not usually exist, it is in many problems possible to find a statistic that is sufficiently close to exactly ancillary that (8.12) remains useful.

Typically $c(\theta, a) = \sqrt{(2\pi)^{-d}}\{1 + O(n^{-1})\}$, and (8.12) approximates the exact conditional density with relative error $O(n^{-3/2})$, although explicit conditions to ensure this are difficult to state precisely.

When $d = 1$, the exponent in (8.12) is equal to $-r^2/2$, where the likelihood root r is given at (8.6); if the exponent is expanded in a Taylor series in $\widehat{\theta}$ we recover the normal approximation to the density of the maximum likelihood estimator from the leading term with variance equal to $j(\widehat{\theta})^{-1}$. This is one indication that standardizing by observed Fisher information is preferable to other first order equivalent alternatives. In the next section we will integrate the p^* approximation to obtain approximations to the distribution function, in the case $d = 1$. The change of variable from $\widehat{\theta}$ to r is the main step in this derivation.

The p^* approximation is exact in location models, and more generally in transformation models, which typically permit a reduction in dimension by conditioning on an ancillary statistic. In these settings the p^* formula generalizes a result of Fisher (1934) for conditional inference in location models. In full exponential families, the p^* approximation can be obtained by a simple change of variable from the saddlepoint approximation to the density of the minimal sufficient statistic. A completely general derivation of the p^* approximation is quite difficult, as it is necessary to find the transformation from y to $(\widehat{\theta}, a)$ and to specify some suitable approximately ancillary statistic a. In the setting of curved exponential families this can be done by embedding the curved exponential family in a full exponential family; see the bibliographic notes.

8.4.2 The tangent exponential model

The p^* approximation gives the density of the maximum likelihood estimator at each point in its sample space, which is why it is necessary to express $\ell(\theta; y)$ as $\ell(\theta; \widehat{\theta}, a)$. A simpler density approximation is available using the likelihood function at (and near) a fixed value of y, denoted y^0. This density approximation is called a *tangent exponential model* and is expressed as

$$p_{\text{TEM}}(s \mid a; \theta) = c |j(\widehat{\varphi})|^{-1/2} \exp[\ell(\theta; y^0) - \ell(\widehat{\theta}^0; y^0) + \{\varphi(\theta) - \varphi(\widehat{\theta}^0)\}^{\mathsf{T}} s], \qquad (8.14)$$

where $s = s(y) = \partial\ell(\widehat{\theta}^0; y)/\partial\theta$, $\widehat{\theta}^0$ is the maximum likelihood estimate at the fixed point y^0, $\varphi(\theta)$ is a sample space derivative of the log likelihood function, defined below at (8.15), and $j(\widehat{\varphi})$ is the observed Fisher information in the new parametrization. The tangent exponential model is a local exponential family model with sufficient statistic s and canonical parameter φ. It has the same log likelihood function as the original model at the fixed point y^0, where it also has the same first derivative with respect to y. In its dependence on the sample space, (8.14) uses only the likelihood function and its first derivative at y^0.

The canonical parameter φ is defined using a set of n vectors of length d, construction of which is discussed in the next subsection. Assume now that we have such a set of vectors available in the form of an $n \times d$ matrix V, with rows $V_1, \ldots, V_n$. Then the definition of $\varphi(\theta)$ is

$$\varphi^{\mathsf{T}}(\theta) = \varphi^{\mathsf{T}}(\theta; y^0) = \ell_{;V}(\theta; y^0) = \sum_{i=1}^{n} \left.\frac{\partial\ell(\theta; y)}{\partial y_i}\right|_{y=y^0} V_i; \qquad (8.15)$$

the last equality holds when $(y_1, \ldots, y_n)$ are independent.

Illustration: Exponential family model

In a full exponential family with density

$$f(y; \theta) = \exp\{\varphi^{\mathrm{T}}(\theta)y - c(\theta) - h(y)\},$$

the canonical parameter $\varphi(\theta)$ is the derivative of the log likelihood function with respect to the minimal sufficient statistic y. This was the original motivation for the development of the tangent exponential model and its canonical parameter φ. Note however that defining the canonical parameter via differentiation of $\ell(\theta; y)$ does not distinguish between affine transformations of the canonical parameter. In the examples we used the most convenient affine transformation of φ.

Illustration: Location family model

In sampling from the model

$$f(y; \theta) = f_0(y - \theta),$$

with θ scalar, the configuration statistic $(a_1, \ldots, a_n) = (y_1 - \widehat{\theta}, \ldots, y_n - \widehat{\theta})$ is exactly ancillary. As described below, the $n \times 1$ matrix V that specifies these ancillary directions is simply $(1, \ldots, 1)^{\mathrm{T}}$, and thus

$$\varphi(\theta) = \ell_{;V}(\theta; y^0) = \sum_{i=1}^{n} \ell_{;y_i}(\theta; y^0) = -\ell_\theta(\theta; y^0). \tag{8.16}$$

Illustration: A curved exponential family

A multinomial model that arises in studies of genetic linkage has four categories with probabilities $p(\theta) = (2 + \theta, 1 - \theta, 1 - \theta, \theta)/4$, where $0 < \theta < 1$. The log likelihood function is

$$\ell(\theta) = y_1 \log(2 + \theta) + (y_2 + y_3) \log(1 - \theta) + y_4 \log \theta,$$

where $y_1, \ldots, y_4$ are the category counts, with $m = y_1 + \cdots + y_4$ fixed. In this discrete model we have $V = (1, -1, -1, 1)$, giving

$$\varphi(\theta) = \log(2 + \theta) - 2 \log(1 - \theta) + \log \theta;$$

in this case φ does not depend on the fixed point y^0. For a numerical illustration we consider data from Rao (1973, p. 369) with $y_1 = 125$, $y_2 = 18$, $y_3 = 20$ and $y_4 = 34$. The left-hand panel of Figure 8.1 shows the log likelihoods as a function of θ for the original

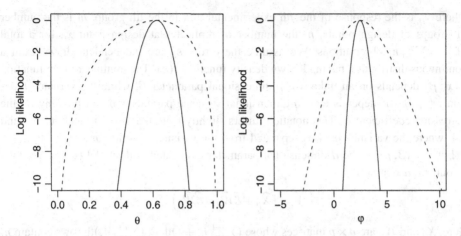

Figure 8.1 Likelihoods for genetic linkage model applied to original data (solid) and to reduced dataset (dashes), in original parametrization (left) and the local reparametrization (right).

data, and for a smaller data set, created by dividing the counts by a factor ten; the latter is more asymmetric because of the decreased sample size. The right-hand panel shows the log likelihoods $\ell(\varphi)$: evidently one effect of the reparametrization is to symmetrize the log likelihood. Figure 8.2 shows that $\varphi(\theta)$ is almost a linear function of θ in the interval $(0.2, 0.8)$, but is otherwise quite nonlinear; this accounts for the symmetrizing effect.

Illustration: Nonlinear regression model

Consider the nonlinear model

$$y_{ij} = \mu(x_i; \beta) + w(x_i; \beta, \rho)\varepsilon_{ij}, \quad i = 1, \ldots, m, \quad j = 1, \ldots, n_i, \qquad (8.17)$$

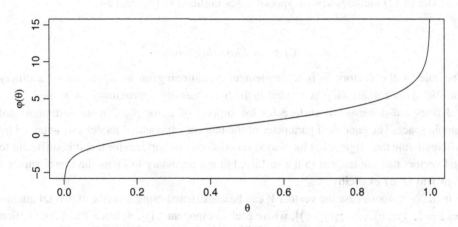

Figure 8.2 Local reparametrization $\varphi(\theta)$ for the genetic linkage model.

where y_{ij} is the response of the jth experimental unit in the ith group, m is the number of groups or design points, n_i the number of replicates at design point x_i, for a total of $n = \sum_{i=1}^{m} n_i$ observations. We assume the errors ε_{ij} are independent draws from a continuous distribution having known density function $f(\varepsilon)$. The nonlinear mean function $\mu(x_i; \beta)$ depends on an unknown p-dimensional parameter β, while the variance function $w(x_i; \beta, \rho)^2$ depends on a q-dimensional variance parameter ρ and possibly on the regression coefficient β. This notation differs slightly from that used in Sections 5.4 and 6.4, where the variance σ^2 was separated from the variance function w^2.

Let $\theta = (\beta, \rho)$ be the d-dimensional parameter associated with model (8.17). We take as our $n \times d$ matrix V

$$V = \left(\widehat{X}_\beta + \widehat{\mathcal{E}}\widehat{\mathcal{W}}_\beta \quad \widehat{\mathcal{E}}\widehat{\mathcal{W}}_\rho \right).$$

Here, X_β and $\mathcal{W}_\beta$ are $n \times p$ matrices whose $(\sum_{r=1}^{i-1} n_r + 1)$th to $(\sum_{r=1}^{i} n_r)$th rows contain n_i replicates of the gradients $\partial \mu(x_i; \beta)/\partial \beta^{\mathrm{T}}$ and $\partial w(x_i; \beta; \rho)/\partial \beta^{\mathrm{T}}$ of the mean and variance functions with respect to β, $\mathcal{W}_\rho$ is an $n \times q$ matrix whose $(\sum_{r=1}^{i-1} n_r + 1)$th to $(\sum_{r=1}^{i} n_r)$th rows are given by $\partial w(x_i; \beta, \rho)/\partial \rho^{\mathrm{T}}$, and $\mathcal{E}$ is a $n \times n$ diagonal matrix whose elements are

$$\varepsilon_{11} = \{y_{11} - \mu(x_1; \beta)\}/w(x_1; \beta, \rho), \ldots, \varepsilon_{mn_m} = \{y_{mn_m} - \mu(x_m; \beta)\}/w(x_m; \beta, \rho).$$

This gives

$$\varphi(\theta)^{\mathrm{T}} = -g_1^{\mathrm{T}} \mathcal{W}^{-1} \left(\widehat{X}_\beta + \widehat{\mathcal{E}}\widehat{\mathcal{W}}_\beta \quad \widehat{\mathcal{E}}\widehat{\mathcal{W}}_\rho \right), \tag{8.18}$$

where $g_1^{\mathrm{T}} = \{g'(\varepsilon_{11}), \ldots, g'(\varepsilon_{mn_m})\}$, with $g(\cdot) = -\log f(\cdot)$ and the prime denoting differentiation, and $\mathcal{W}$ is an $n \times n$ diagonal matrix with elements

$$\mathcal{W}(\beta, \rho) = \mathrm{diag}\{\ldots, \underbrace{w(x_i; \beta, \rho), \ldots, w(x_i; \beta, \rho)}_{n_i}, \ldots\}.$$

Model (8.17) includes several special cases outlined in Problem 54.

8.4.3 Ancillary directions

The role of the vectors V is to implement conditioning on an approximate ancillary statistic. This conditioning is needed in both the density approximations p^* and p_{TEM}: it defines a d-dimensional surface for the support of p^* or p_{TEM} in the n-dimensional sample space. The canonical parameter of the tangent exponential model φ is obtained by differentiating the original log likelihood function on this surface: for this it is sufficient to find vectors that are tangent to the surface, but not necessary to know the transformation from y to (s, a) explicitly.

In the continuous case the vectors V can be constructed using a vector of pivotal quantities $z = \{z_1(y_1, \theta), \ldots, z_n(y_n, \theta)\}$, where each component $z_i(y_i, \theta)$ has a fixed distribution under the model. Such a vector always exists in the form of the probability integral transformation $F(y_i; \theta)$, although simpler alternatives may be available. The matrix V is defined from z by

$$V = - \left(\frac{\partial z}{\partial y^{\mathrm{T}}}\right)^{-1} \left(\frac{\partial z}{\partial \theta^{\mathrm{T}}}\right)\Bigg|_{(y^0, \hat{\theta}_0)}. \tag{8.19}$$

Illustration: Location-scale model

Suppose that $y_i = \mu + \sigma \varepsilon_i$, where ε_i has a known distribution $f(\varepsilon)$. We use $z_i = \varepsilon_i$ as the pivotal statistic for defining V. Then

$$\frac{\partial z}{\partial y^{\mathrm{T}}} = \sigma^{-1} I_n,$$

and

$$\frac{\partial z}{\partial \theta^{\mathrm{T}}} = -\sigma^{-1} \begin{pmatrix} 1 & (y_1 - \mu)/\sigma \\ \vdots & \\ 1 & (y_n - \mu)/\sigma \end{pmatrix},$$

giving

$$V = \begin{pmatrix} 1 & \hat{z}_1 \\ 1 & \hat{z}_2 \\ \vdots & \\ 1 & \hat{z}_n \end{pmatrix},$$

where $\hat{z}_i = (y_i - \hat{\mu})/\hat{\sigma}$ is computed using the maximum likelihood estimate and observed data.

In this model $\hat{z}$ is the maximal ancillary statistic and V reproduces the tangent vectors to that ancillary. The canonical parameter is

$$\varphi(\theta) = \varphi(\theta; y) = \begin{pmatrix} \dfrac{1}{\sigma} \sum_{i=1}^{n} g'\left(\dfrac{y_i - \mu}{\sigma}\right) \\ \dfrac{1}{\sigma} \sum_{i=1}^{n} g'\left(\dfrac{y_i - \mu}{\sigma}\right) \dfrac{y_i - \hat{\mu}}{\hat{\sigma}} \end{pmatrix}, \tag{8.20}$$

where $g(\cdot) = -\log f(\cdot)$. In the normal case, $g(x) = x^2/2$, and $\varphi^{\mathrm{T}} = (n(\hat{\mu} - \mu)/\sigma^2, n\hat{\sigma}/\sigma^2)$ is an affine transformation of $(\mu/\sigma^2, 1/\sigma^2)$, the usual expression of the canonical parameter.

Illustration: An exponential family model

Suppose y_i follows a gamma distribution with shape parameter β and mean μ:

$$f(y_i; \beta, \mu) = \frac{1}{\Gamma(\beta)} \frac{\beta}{\mu} \left(\frac{\beta y_i}{\mu}\right)^{\beta - 1} \exp\left(-\frac{\beta y_i}{\mu}\right), \qquad y_i \geq 0, \quad \mu, \beta > 0.$$

For a pivotal statistic for y_i we use $z_i = F(y_i; \beta, \mu)$, the cumulative distribution function for the gamma density. The ith row of V is

$$(c(\widehat{\beta}, \widehat{\mu}; y_i), y_i/\widehat{\mu}),$$

evaluated at the maximum likelihood estimate and observed data, and

$$c(\beta, \mu; y) = -\frac{\partial F(y; \beta, \mu)/\partial \beta}{f(y; \beta, \mu)}$$

and

$$\varphi(\beta, \mu) = \begin{pmatrix} \beta \sum_{i=1}^{n} (1/y_i - 1/\mu) c(\widehat{\beta}, \widehat{\mu}; y_i) \\ \beta\{(n/\widehat{\mu}) - n/\mu\} \end{pmatrix}$$

is again an affine transformation of the canonical parameters, here $(\beta, \beta/\mu)$.

If y has a discrete distribution the construction of V using pivotal quantities is not usually available, and special arguments are needed; see Section 8.5.4 and Problem 55.

Illustration: Nonlinear regression model

Consider again model (8.17). As in the location-scale model we use $z_{ij} = \varepsilon_{ij} = \{y_{ij} - \mu(x_i; \beta)\}/w(x_i; \beta, \rho)$ as the pivotal statistic for defining the matrix V, which has $n = \sum_{i=1}^{m} n_i$ rows and $d = \dim(\beta, \rho)$ columns. Then

$$\frac{\partial z}{\partial y^{\mathrm{T}}} = \mathcal{W}^{-1}$$

is an $n \times n$ diagonal matrix with elements $w^{-1}(x_1; \beta, \rho), \ldots, w^{-1}(x_m; \beta, \rho)$ in turn replicated as many times as there are observations in the ith group, and

$$\frac{\partial z}{\partial \theta^{\mathrm{T}}} = -\left(\mathcal{W}^{-1}(X_\beta + \mathcal{E}\mathcal{W}_\beta) \quad \mathcal{W}^{-1}\mathcal{E}\mathcal{W}_\rho \right)$$

is an $n \times d$ matrix, with X_β, $\mathcal{W}_\beta$, $\mathcal{W}_\rho$ and $\mathcal{E}$ defined as on page 144. This yields

$$V = \left(\widehat{X}_\beta + \widehat{\mathcal{E}\mathcal{W}}_\beta \quad \widehat{\mathcal{E}\mathcal{W}}_\rho \right).$$

It is left as an exercise to verify that when the variance function is constant we get $V = (\widehat{X}_\beta \quad \widehat{\mathcal{E}})$, that in a linear regression model we have $V = (X + \widehat{\mathcal{E}\mathcal{W}}_\beta \quad \widehat{\mathcal{E}\mathcal{W}}_\rho)$ and that in the classical linear model we have $V = (X \quad \widehat{\mathcal{E}})$, a generalization of (8.20).

8.5 Tail area approximations

8.5.1 Introduction

The density approximations of the previous section are mainly useful for deriving approximations to the cumulative distribution function, as it is this that gives P-values for testing particular values for ψ. In regular problems the P-value as a function of ψ provides confidence limits at any desired confidence. Examples of such P-value functions are plotted in Figures 2.2 and 3.3. In other examples we plot the approximate normal pivot directly as a function of ψ.

The approximations used for nearly all our examples have the forms

$$\Phi^*(r) \equiv \Phi(r) + \phi(r) \left(\frac{1}{r} - \frac{1}{q} \right) \tag{8.21}$$

and

$$\Phi(r^*) \equiv \Phi \left(r + \frac{1}{r} \log \frac{q}{r} \right) \tag{8.22}$$

for suitably defined r and q, where $\phi(\cdot)$ is the density function of the standard normal distribution and $\Phi(\cdot)$ is the corresponding distribution function. The first of these is known as the *Lugannani–Rice* approximation, and the second as *Barndorff-Nielsen's* approximation; see Appendix A.2 and the bibliographic notes.

Assuming for the moment that r and q are related by

$$r = q + \frac{A}{n^{1/2}} q^2 + \frac{B}{n} q^3 + O(n^{-3/2}), \tag{8.23}$$

where A and B are $O(1)$, it follows that

$$\frac{1}{r} \log \left(\frac{r}{q} \right) = \frac{A}{n^{1/2}} + \frac{B - 3A^2/2}{n} q + O(n^{-3/2}),$$

and

$$\frac{1}{r} - \frac{1}{q} = -\frac{A}{n^{1/2}} - \frac{B - A^2}{n} q + O(n^{-3/2}).$$

Then by expanding (8.22) about r we see that (8.21) and (8.22) are equivalent to $O(n^{-1})$; that is, ignoring terms of $O(n^{-3/2})$ and higher. Approximation (8.22) is often more convenient for diagnostic purposes, and has the advantage that it is guaranteed to lie between 0 and 1. Approximation (8.21) is sometimes more accurate, although we are not aware of any general comparison of their numerical performance.

8.5.2 No nuisance parameters

Integrating p^*

We will sketch the derivation of (8.21) and (8.22) from the p^* approximation (8.13) for θ a scalar parameter. We wish to find

$$\int_{-\infty}^{\widehat{\theta}^0} f(\widehat{\theta} \mid a; \theta) \mathrm{d}\widehat{\theta} \doteq \int_{-\infty}^{\widehat{\theta}^0} c |j(\widehat{\theta})|^{1/2} \exp\{\ell(\theta) - \ell(\widehat{\theta})\} \mathrm{d}\widehat{\theta}, \tag{8.24}$$

where $\widehat{\theta}^0$ is the observed data point, and (8.24) gives the (left tail) P-value for testing the hypothesized value θ. Here, and below, we use $\doteq$ to mean 'is approximately equal to'. We will assume that $\widehat{\theta}^0 < \theta$. Let

$$\frac{1}{2} r^2 = \ell(\widehat{\theta}) - \ell(\theta)$$

so that

$$r \, \mathrm{d}r = \{\ell_{;\widehat{\theta}}(\widehat{\theta}) - \ell_{;\widehat{\theta}}(\theta)\} \mathrm{d}\widehat{\theta}.$$

Then (8.24) becomes

$$\int_{-\infty}^{r^0} c \exp(-r^2/2)(r/q) \mathrm{d}r, \tag{8.25}$$

with

$$r = \mathrm{sign}(\widehat{\theta} - \theta)[2\{\ell(\widehat{\theta}) - \ell(\theta)\}]^{1/2}, \tag{8.26}$$

$$q = \{\ell_{;\widehat{\theta}}(\widehat{\theta}; \widehat{\theta}, a) - \ell_{;\widehat{\theta}}(\theta; \widehat{\theta}, a)\} j(\widehat{\theta})^{-1/2}. \tag{8.27}$$

Since c is the normalizing constant in the p^* approximation (8.12), it follows from (8.25) that $c = \sqrt{(2\pi)}\{1 + O(n^{-1})\}$ (see Problem 56), and then an application of the argument in Appendix A.2 gives the approximations

$$\Pr(\widehat{\theta} \le \widehat{\theta}^0; \theta) \doteq \Phi(r^0) + \phi(r^0)\left(\frac{1}{r^0} - \frac{1}{q^0}\right),$$

$$\doteq \Phi\left(r^0 + \frac{1}{r^0} \log \frac{q^0}{r^0}\right), \tag{8.28}$$

where r^0 and q^0 are defined by (8.26) and (8.27) with $\widehat{\theta}$ replaced by $\widehat{\theta}^0$. The assumption that we can write r as an expansion of q in the form (8.23) follows from an expansion of $\ell(\widehat{\theta}; \widehat{\theta}, a) - \ell(\theta; \widehat{\theta}, a)$. We are also assuming enough regularity to ensure that the transformation from $\widehat{\theta}$ to r is monotone, and that the requisite Taylor expansions are valid. Both r^0 and q^0 are functions of $\widehat{\theta}^0$ and θ, that is, of the observed maximum likelihood estimate and the hypothesized value θ. When it is clear from the context we use $\widehat{\theta}$ for the observed maximum likelihood estimate $\widehat{\theta}^0$.

In the case that $\widehat{\theta}^0 > \theta$ we use the same argument to approximate the corresponding integral from $\widehat{\theta}^0$ up to infinity; this detail has been incorporated into (8.26) and (8.27) by ensuring that r and q have the same sign.

There is a discontinuity in (8.28) at the point $\theta = \widehat{\theta}^0$, where both r and q are zero, although the approximations are continuous at that point. In practice this can cause numerical instability for a range of θ values near $\widehat{\theta}^0$. In all examples where it was needed we used the tail area approximation in the range for θ where it is stable, and interpolated through the discontinuity. This is implemented in the cond and marg packages in the hoa package bundle by replacing the approximation for all θ values in an interval containing $\widehat{\theta}$ by a high-degree polynomial; see Section 9.3.

We see at (8.25) that the density for r implied by the p^* approximation is

$$(r/q)\phi(r)\left\{1+D/n+O(n^{-3/2})\right\}.$$

Assuming again expansion (8.23) for r in terms of q, we have

$$q = r - \frac{A}{n^{1/2}}r^2 + \frac{2A^2 - B}{n}r^3 + O(n^{-3/2}),$$

$$D = A^2 - B,$$

$$\mathrm{E}(r) = \frac{A}{n^{1/2}} + O(n^{-3/2}),$$

$$\mathrm{var}(r) = 1 + \frac{1}{n}(2B - 3A^2) + O(n^{-2}),$$

and hence

$$r^* = \frac{r - A/n^{1/2}}{\{1 + (2B - 3A^2)/n\}^{1/2}} + O(n^{-3/2}) = \frac{r - \mathrm{E}(r)}{\{\mathrm{var}(r)\}^{1/2}} + O(n^{-3/2}), \qquad (8.29)$$

showing that to the same order of approximation, both (8.21) and (8.22) are equivalent to the normal approximation to the mean and variance corrected version of r. This normal approximation is the scalar version of the Bartlett correction of w discussed in Section 8.8.2. It has however proved to be numerically much less accurate than $\Phi(r^*)$ in most applications.

Integrating p_{TEM}

In order to carry out the integration above, we need only the first derivative of ℓ in the sample space with the ancillary statistic a held fixed. For this it is sufficient to know the tangent vectors to the surface on the sample space determined by fixed a, and not necessary to know the form of the ancillary or the transformation from y to $(\widehat{\theta}, a)$ explicitly. This is the motivation for the development of the tangent exponential model (8.14). Using p_{TEM} to derive the tail area approximation gives the resulting expression for q^0 as

$$q^0 = (\widehat{\varphi}^0 - \varphi)j_{\varphi\varphi}(\widehat{\varphi}^0)^{1/2}, \qquad (8.30)$$

where φ, defined at (8.15), is the canonical parameter of the tangent exponential model. Both approximations in (8.28) hold with r^0 unchanged and this q^0. An equivalent expression for q^0 which compares more directly with (8.27) is

$$q^0 = j(\widehat{\theta}^0)^{1/2} |\ell_{\theta;V}(\widehat{\theta})|^{-1} \{\ell_{;V}(\widehat{\theta}^0) - \ell_{;V}(\theta)\}. \tag{8.31}$$

If V is tangent to the statistic a used in (8.27), then these expressions are identical.

Illustration: Exponential family model

As shown in the previous subsection, for a sample of size n from the scalar parameter model

$$f(y; \theta) = \exp\{\theta y - c(\theta) - d(y)\},$$

the canonical parameter φ defined by $\ell_{;y}(\theta; y^0)$ is equivalent to θ, and thus q from (8.30) is simply the standardized maximum likelihood estimate for the canonical parameter. Alternatively, we can use the p^* approximation to the density of the maximum likelihood estimate, as the likelihood function based on the sample can be expressed as a function of $\widehat{\theta}$ and θ, with no need for a complementary ancillary statistic a. That is,

$$\ell(\theta; \widehat{\theta}) = \theta n c'(\widehat{\theta}) - n c(\theta),$$

from which we see that $\ell_{;\widehat{\theta}}(\theta) = \theta n c''(\widehat{\theta})$; thus (8.27) becomes

$$q = t = j(\widehat{\theta})^{1/2}(\widehat{\theta} - \theta).$$

For a numerical illustration see Section 2.2.

Illustration: Location model

In a one parameter location model, $f(y; \theta) = f(y - \theta)$, the maximum likelihood estimate based on a sample $(y_1, \ldots, y_n)$ is the solution of the equation

$$\sum g'(y_i; \widehat{\theta}) = 0,$$

where $g(\cdot) = -\log f(\cdot)$, $g'(y_i; \theta) = \partial g(y_i; \theta)/\partial\theta$, and the configuration vector

$$a = (a_1, \ldots, a_n) = (y_1 - \widehat{\theta}, \ldots, y_n - \widehat{\theta})$$

is exactly ancillary. The transformation from $(y_1, \ldots, y_n)$ to $(\widehat{\theta}, a)$ is one-to-one, as a has only $n-1$ free dimensions, and we can write

$$\ell(\theta; \widehat{\theta}, a) = -\sum g(a_i + \widehat{\theta} - \theta),$$

from which we see that $\ell_{;\widehat{\theta}} = -\ell_{\theta}$, and (8.27) becomes

$$q = s = j(\widehat{\theta})^{-1/2}\ell_{\theta}(\theta),$$

the standardized score function. Using the tangent exponential model approach as in Section 8.4.3, V is simply a vector of ones, and we get the same expression for q. For a numerical example see Section 3.2.

8.5.3 Nuisance parameters

Expressions for q

We now present the formulae for the general P-value approximations in the presence of nuisance parameters, using the notation of Section 8.2. The P-value expressions are identical to (8.21) and (8.22), with the likelihood root computed from the profile log likelihood:

$$r = r_{\mathrm{p}} = \mathrm{sign}(\widehat{\psi} - \psi)[2\{\ell_{\mathrm{p}}(\widehat{\psi}) - \ell_{\mathrm{p}}(\psi)\}]^{1/2}.$$

Two basic expressions for q are available, as extensions of (8.27) and (8.31):

$$q_1 = \frac{|\ell_{;\widehat{\theta}}(\theta) - \ell_{;\widehat{\theta}}(\theta_{\psi}) \quad \ell_{\lambda;\widehat{\theta}}(\theta_{\psi})|}{|\ell_{\theta;\widehat{\theta}}(\widehat{\theta})|} \left\{ \frac{|j_{\theta\theta}(\widehat{\theta})|}{|j_{\lambda\lambda}(\widehat{\theta}_{\psi})|} \right\}^{1/2} \tag{8.32}$$

and

$$q_2 = \frac{|\varphi(\widehat{\theta}) - \varphi(\widehat{\theta}_{\psi}) \quad \varphi_{\lambda}(\widehat{\theta}_{\psi})|}{|\varphi_{\theta}(\widehat{\theta})|} \left\{ \frac{|j_{\theta\theta}(\widehat{\theta})|}{|j_{\lambda\lambda}(\widehat{\theta}_{\psi})|} \right\}^{1/2}. \tag{8.33}$$

The $d \times d$ matrices appearing in the numerators of the first terms of both q_1 and q_2 consist of a column vector formed using sample space derivatives, and a $d \times (d-1)$ matrix of mixed derivatives. These are in general rather unwieldy, and in Section 8.6 we give explicit forms as illustrations. A version that may be more convenient for some computations can be obtained by first noting that the score equation

$$\ell_{\theta}(\widehat{\theta}; \widehat{\theta}, a) = 0$$

can be differentiated with respect to $\widehat{\theta}$ to show that

$$\ell_{\theta;\widehat{\theta}}(\widehat{\theta}; \widehat{\theta}, a) = j(\widehat{\theta}),$$

and the determinant in the numerator of (8.32) can be computed by writing the matrix in a partitioned form with a scalar entry in the upper left corner, and a $(d-1) \times (d-1)$ submatrix in the lower right corner. The resulting expression is

$$q_1 = [\ell_{;\widehat{\psi}}(\widehat{\theta}) - \ell_{;\widehat{\psi}}(\widehat{\theta}_{\psi}) - \ell_{\lambda;\widehat{\psi}}(\widehat{\theta}_{\psi})\ell_{\lambda;\widehat{\lambda}}^{-1}(\widehat{\theta}_{\psi})\{\ell_{;\widehat{\lambda}}(\widehat{\theta}) - \ell_{;\widehat{\lambda}}(\widehat{\theta}_{\psi})\}]$$
$$\times |j(\widehat{\theta})|^{-1/2}|j_{\lambda\lambda}(\widehat{\theta}_{\psi})|^{-1/2}|\ell_{\lambda;\widehat{\lambda}}(\widehat{\theta}_{\psi})|.$$

An equivalent expression for q_2 as a type of Wald statistic, as at (8.30), is computed from the canonical parameter of the tangent exponential model, $\varphi(\theta)$, as

$$\chi(\theta) = \frac{\psi_{\varphi^{\mathrm{T}}}(\widehat{\theta}_\psi)}{|\psi_{\varphi^{\mathrm{T}}}(\widehat{\theta}_\psi)|}\varphi(\theta).$$

This scalar is the projection of φ onto the direction taken by the derivative of the interest parameter ψ with respect to φ, evaluated at the constrained maximum likelihood estimate $\widehat{\theta}_\psi = (\psi, \widehat{\lambda}_\psi)$. The corresponding expression for q is

$$q_2 = \mathrm{sign}(\widehat{\psi} - \psi)\,|\chi(\widehat{\theta}) - \chi(\widehat{\theta}_\psi)|\left\{\frac{|j_{(\theta\theta)}(\widehat{\theta})|}{|j_{(\lambda\lambda)}(\widehat{\theta}_\psi)|}\right\}^{1/2}, \tag{8.34}$$

where the two determinants in the ratio are those appearing in (8.32) and (8.33), but recalibrated in terms of the φ parametrization:

$$|j_{(\theta\theta)}(\widehat{\theta})| = |j_{\theta\theta}(\widehat{\theta})|\,|\varphi_\theta(\widehat{\theta})|^{-2}, \quad |j_{(\lambda\lambda)}(\widehat{\theta}_\psi)| = |j_{\lambda\lambda}(\widehat{\theta}_\psi)|\,|\varphi_\lambda(\widehat{\theta}_\psi)^{\mathrm{T}}\varphi_\lambda(\widehat{\theta}_\psi)|^{-1}.$$

Once the log likelihood and φ as functions of $\theta = (\psi, \lambda)$ are available, numerical differentation yields the necessary maximum likelihood estimates, information matrices and derivatives φ_θ φ_λ, and ψ_φ, the latter as a column of the matrix inverse of φ_θ. The calculation of χ and the recalibrated information matrices is carried out in detail for a simple special case in Section 3.5.

A similar discussion applies when the parameter of interest is given only implicitly as $\psi(\theta)$. In this case the overall maximum likelihood estimate is $\widehat{\psi} = \psi(\widehat{\theta})$, with estimated variance $\sigma^2(\widehat{\theta})$, where $\sigma^2(\theta) = \psi_\theta(\theta)^{\mathrm{T}}j_{\theta\theta}^{-1}(\theta)\psi_\theta(\theta)$. The constrained estimate $\widehat{\theta}_\psi$ may be obtained by maximizing the Lagrangian

$$\tilde{\ell}(\theta, \alpha) = \ell(\theta) + \alpha\left\{\psi(\theta) - \psi\right\}$$

with respect to both θ and the Lagrange multiplier α. Introduction of the constraint $\psi(\theta) = \psi$ makes numerical maximization awkward, and it may be necessary to add a term $-C^2\left\{\psi(\theta) - \psi\right\}^2$ to the Lagrangian to ensure that the maximum is obtained. In this setting,

$$q_2 = \mathrm{sign}(\widehat{\psi} - \psi)\frac{|\overline{\chi}(\widehat{\theta}) - \overline{\chi}(\widehat{\theta}_\psi)|}{\widehat{\sigma}_\chi(\varphi)}, \tag{8.35}$$

where

$$\overline{\chi}(\theta) = \psi_\varphi(\widehat{\theta}_\psi)\varphi(\theta) = \psi_\theta(\widehat{\theta}_\psi)\{\varphi_\theta(\widehat{\theta}_\psi)^{\mathrm{T}}\}^{-1}\varphi(\theta),$$

$$\widehat{\sigma}_\chi^2(\varphi) = \left\{\psi_\theta(\widehat{\theta}_\psi)j(\widehat{\theta}_\psi)^{-1}\psi_\theta(\widehat{\theta}_\psi)^{\mathrm{T}}\right\}\frac{|j_{(\theta\theta)}(\widehat{\theta}_\psi)|}{|j_{(\theta\theta)}(\widehat{\theta})|}.$$

Direct specification of the ancillary statistic can also be avoided if p^* approximations accurate to $O(n^{-1})$ instead of $O(n^{-3/2})$ are used; see the bibliographic notes. Among several asymptotically equivalent formulations, the approximation to q_1 in (8.32) due to Skovgaard (1996), which avoids the use of sample space derivatives, seems to be the most promising at the time of writing. The numerator of q_1 may be expressed as (Problem 63)

$$|\ell_{\theta;\widehat{a}}(\widehat{\theta}_\psi)|[\{\ell_{;\widehat{a}}(\widehat{\theta}) - \ell_{;\widehat{a}}(\widehat{\theta}_\psi)\}^{\mathrm{T}}\{\ell_{\theta;\widehat{a}}(\widehat{\theta}_\psi)\}^{-1}]_1, \tag{8.36}$$

where the second factor is the first component of the vector in square brackets: this presupposes that ψ is the first component of θ. Skovgaard (1996) developed moment-based approximations to the sample space derivatives:

$$\ell_{;\widehat{a}}(\widehat{\theta}) - \ell_{;\widehat{a}}(\widehat{\theta}_\psi) \doteq i(\widehat{\theta})^{-1} j(\widehat{\theta}) Q(\widehat{\theta}, \widehat{\theta}_\psi), \tag{8.37}$$

$$\ell_{\theta;\widehat{a}}(\widehat{\theta}_\psi)^{\mathrm{T}} \doteq i(\widehat{\theta})^{-1} j(\widehat{\theta}) S(\widehat{\theta}, \widehat{\theta}_\psi),$$

where

$$S(\theta_1, \theta_2) = \mathrm{cov}_{\theta_1}\{\ell_\theta(\theta_1), \ell_\theta(\theta_2)\}, \tag{8.38}$$

$$Q(\theta_1, \theta_2) = \mathrm{cov}_{\theta_1}\{\ell_\theta(\theta_1), \ell(\theta_1) - \ell(\theta_2)\}. \tag{8.39}$$

The resulting approximation is

$$q_1 \doteq [\widehat{S}^{-1}\widehat{Q}]_1 |j(\widehat{\theta})|^{1/2} |i^{-1}(\widehat{\theta})| |\widehat{S}| |j_{\lambda\lambda}(\widehat{\theta}_\psi)|^{-1/2}, \tag{8.40}$$

where $\widehat{S} = S(\widehat{\theta}, \widehat{\theta}_\psi)$, $\widehat{Q} = Q(\widehat{\theta}, \widehat{\theta}_\psi)$.

An advantage of this formulation is that the covariances S and Q can be computed in many problems, and there is no need to specify either an approximate ancillary or a pivotal statistic for computing ancillary directions. A disadvantage is that the tail area approximation using (8.40) has relative error $O(n^{-1})$, rather than $O(n^{-3/2})$. Explicit expressions for S and Q are readily computed in nonlinear regression with normal errors of non-constant variance, and approximations to the tail area using (8.40) in $\Phi(r^*)$ or $\Phi^*(r)$ are implemented in the `nlreg` package of the `hoa` package bundle.

Modifications to profile likelihood

In most of the examples we compute the first order approximations of Section 8.3, as well as the higher order Barndorff–Nielsen and Lugannani–Rice approximations of this section, (8.21) and (8.22). Some insight into the difference between the two can sometimes be gained by plotting both the profile log likelihood $\ell_p(\psi)$ and a modification of the profile log likelihood suggested by the higher order approximations. This plot is built into the packages `cond` and `marg` in the `hoa` bundle.

There are two general forms of modified profile likelihood, one derived using the p^* approximation and the tail area approximation using q_1, (8.32), and the other derived using the tangent exponential model and the tail area approximation using q_2, (8.33). They both simplify considerably when the model is a linear exponential family or a linear non-normal regression, as is indicated in Section 8.6 below.

The first general form is given by

$$\ell_{\mathrm{m}}(\psi) = \ell_{\mathrm{p}}(\psi) + \frac{1}{2}\log|j_{\lambda\lambda}(\widehat{\theta}_\psi)| - \log|\ell_{\lambda;\widehat{\lambda}}(\psi, \widehat{\lambda}_\psi; \widehat{\psi}, \widehat{\lambda}, a)|, \tag{8.41}$$

where

$$\ell_{\lambda;\widehat{\lambda}} = \frac{\partial^2 \ell(\psi, \lambda; \widehat{\psi}, \widehat{\lambda}, a)}{\partial \widehat{\lambda} \partial \lambda^\mathsf{T}}. \tag{8.42}$$

The analogous modified profile log likelihood derived from the tangent exponential model is

$$\ell_{\mathrm{F}}(\psi) = \ell_{\mathrm{p}}(\psi) + \frac{1}{2}\log|j_{\lambda\lambda}(\widehat{\theta}_\psi)| - \frac{1}{2}\log|\varphi_\lambda(\widehat{\theta}_\psi)^\mathsf{T} j_{\varphi\varphi}(\widehat{\theta})\varphi_\lambda(\widehat{\theta}_\psi)|. \tag{8.43}$$

Plots of the modified profile log likelihood $\ell_{\mathrm{m}}(\psi)$ are part of the plotting output from the `cond` and `marg` packages in the `hoa` package bundle, as in these cases the ancillary statistic can be specified explicitly and the sample space derivatives readily calculated. In examples such as Section 4.4 where we used the tangent exponential model and the related tail area approximation we computed the modified likelihood with code specific to the example considered.

8.5.4 Discrete random variables

In Chapter 4 and in Sections 3.3 and 3.4 we applied the higher order approximations (8.21) and (8.22) to Poisson and binomial models. The exact distribution function for the maximum likelihood estimator or the likelihood root has jumps, typically of size $O(n^{-1/2})$, at the lattice points for the model. We view the approximations in this setting as approximations to a smoothed version of the true distribution function, obtained by spreading the mass at each jump over the interval around the lattice point. The simplest way to achieve this is to approximate the *mid*-P-value: $\Pr(y < y^0) + (\frac{1}{2})\Pr(y = y^0)$.

An alternative is to incorporate continuity correction into r and q. In Section 3.3 we did this by computing r and q using $y^0 + \frac{1}{2}$ in place of y^0, and obtained an approximation which was very accurate at the jump points of the distribution. This accuracy is illustrated in Figure 3.2 and borne out in simulations of binomial and Poisson models with a scalar parameter. For logistic regression a slightly different continuity correction, suggested by detailed examination of the underlying saddlepoint approximation, is incorporated into the `cond` package. However, conditioning has effects on the lattice structure that are difficult to assess precisely. As a result we usually do not incorporate continuity correction and use mid-P-values instead.

A further difficulty in discrete models arises in determining the canonical parameter φ of the tangent exponential model by sample space differentiation. In full exponential family models this difficulty does not arise because φ is the usual canonical parameter of that model, but in more general models the calculation described at (2.15) is needed; see Problem 55.

8.6 Tail area expressions for special cases

8.6.1 Linear exponential family

We suppose that $y = (y_1, \ldots, y_n)$ is a sample of independent and identically distributed observations from a full exponential family model

$$f(y_i; \psi, \lambda) = \exp\{\psi u(y_i) + \lambda^{\mathrm{T}} v(y_i) - c(\psi, \lambda) + h(y_i)\}, \qquad (8.44)$$

where $u(y_i)$ is scalar and $v(y_i) = \{v_1(y_i), \ldots, v_{d-1}(y_i)\}$. We write $(u, v) = (\sum u(y_i), \sum v(y_i))$ for the minimal sufficient statistic, with u and v associated with ψ and λ respectively. Since $\varphi(\theta)$ is equivalent to θ, we have

$$\varphi(\widehat{\theta}) - \varphi(\widehat{\theta}_\psi) = \begin{pmatrix} \widehat{\psi} - \psi \\ \widehat{\lambda} - \widehat{\lambda}_\psi \end{pmatrix}$$

and $\varphi_{\theta^{\mathrm{T}}}(\theta) = I_d$, so approximations (8.21) and (8.22) use

$$r = \mathrm{sign}(\widehat{\psi} - \psi)[2\{\ell_{\mathrm{p}}(\widehat{\psi}) - \ell_{\mathrm{p}}(\psi)\}]^{1/2},$$
$$q_2 = (\widehat{\psi} - \psi)|j(\widehat{\theta})|^{1/2} / |j_{\lambda\lambda}(\widehat{\theta}_\psi)|^{-1/2}$$
$$= (\widehat{\psi} - \psi) j_{\mathrm{p}}(\widehat{\psi})^{1/2} \rho(\psi, \widehat{\psi}), \qquad (8.45)$$

where

$$\rho(\psi, \widehat{\psi}) = \left\{ |j_{\lambda\lambda}(\widehat{\psi}, \widehat{\lambda})| / |j_{\lambda\lambda}(\psi, \widehat{\lambda}_\psi)| \right\}^{1/2}; \qquad (8.46)$$

the second version of q_2 makes use of (8.4). As in the one parameter exponential family, q is a standardized maximum likelihood estimate, but now modified by the adjustment ρ that arises in the elimination of the nuisance parameters. By extending the p^* approach given in the exponential family illustration of Section 8.5.3 we can verify that $q_1 = q_2$; see Problem 58.

The difference between r^* and r is $r^{-1} \log(q_2/r)$, and we can see from (8.45) that this can be expressed as $r^{-1} \log(t/r) + r^{-1} \log \rho$, where t is the Wald statistic for ψ; see also (2.9). These are the two components r_{INF} and r_{NP} mentioned at (2.10). The form of ρ suggests that the component r_{NP} corresponds to the adjustment for nuisance parameters, although this is difficult to make precise.

In exponential families in which the parameter of interest is a linear function of the canonical parameter, the nuisance parameter can be exactly eliminated by conditioning. It can be shown that the Lugannani–Rice approximation $\Phi^*(r)$ and the Barndorff-Nielsen approximation $\Phi(r^*)$ implement this conditional inference in the linear exponential family model. First we note that

$$f_c(u \mid v; \psi) = \frac{f(u, v; \psi, \lambda)}{f(v; \psi, \lambda)} = \exp\{\psi u - k_v(\psi) - h_v(u)\}$$

is again an exponential family model, with log likelihood function

$$\ell_c(\psi) = \psi u - k_v(\psi).$$

Applying the saddlepoint approximation to the joint distribution of (u, v) and separately to the marginal distribution of v gives a p^*-type approximation to the conditional density of u given v. Integrating the p^* approximation to this conditional density gives approximations of the form (8.21) and (8.22) with r and q as above; see Problem 59.

It can also be shown that

$$f_c(u \mid v; \psi) = \exp\{\ell_m(\psi)\}\{1 + O(n^{-1})\},$$

where $\ell_m(\psi)$, defined at (8.41), simplifies in this setting to

$$\ell_m(\psi) = \ell_p(\psi) + \frac{1}{2} \log |j_{\lambda\lambda}(\psi, \widehat{\lambda}_\psi)|. \tag{8.47}$$

We can obtain an equivalent tail area approximation with $\ell_m(\psi)$ by taking

$$r = \text{sign}(q)[2\{\ell_m(\widehat{\psi}_m) - \ell_m(\psi)\}]^{1/2},$$
$$q = j_m(\widehat{\psi}_m)^{1/2}(\widehat{\psi}_m - \psi).$$

This version has the same relative error, $O(n^{-3/2})$, as that based on (8.45). In our examples of earlier chapters we often plot ℓ_m, but inference is usually based on the version of r^* constructed from ℓ_p, (8.45); the cell phone example of Section 4.3 is an exception. Approximations using (8.45) are the basis of the `cond` package in the `hoa` package bundle, and summaries using (8.47) are also available. See Problem 59 and the bibliographic notes.

8.6.2 Non-normal linear regression

This generalizes the location model illustration of Section 8.5.2. Suppose

$$y_i = x_i^T \beta + \sigma \varepsilon_i, \tag{8.48}$$

where x_i is a known vector of covariates of length p, and the errors ε_i are independent and identically distributed from a known distribution $f(\varepsilon)$. We first assume that the parameter of interest is a component of the regression parameter β, say β_1. We will use this example to illustrate the general forms for q given in Section 8.5.3, at (8.32) and (8.33).

The quantities $z_i = \varepsilon_i = (y_i - x_i^{\mathrm{T}}\beta)/\sigma$, for $i = 1, \ldots, n$, are the pivots used for defining the tangent vectors V, which can be obtained as outlined in Section 8.4.3 at (8.19) or as a special case of the nonlinear regression model:

$$V = \begin{pmatrix} x_1^{\mathrm{T}} & \widehat{\varepsilon}_1 \\ \vdots & \vdots \\ x_n^{\mathrm{T}} & \widehat{\varepsilon}_n \end{pmatrix},$$

where $\widehat{\varepsilon}_i = (y_i - x_i^{\mathrm{T}}\widehat{\beta})/\widehat{\sigma}$ is the ith standardized residual; the vector $\widehat{\varepsilon}$ of residuals is the usual ancillary statistic for this model. Using this we have

$$\varphi(\theta)^{\mathrm{T}} = -\widehat{\sigma}^{-1}\sum_{i=1}^{n} g'(\varepsilon_i)\,(x_i^{\mathrm{T}} \quad \widehat{\varepsilon}_i),$$

where $g(x) = -\log f(x)$ and the prime denotes differentiation.

We present the formulae for q_2 when the parameter of interest is β_1, so the nuisance parameter is $\lambda = (\beta_2, \ldots, \beta_p, \sigma)$. In a departure from our usual notation let $\tilde{\beta}$ and $\tilde{\sigma}$ denote the constrained maximum likelihood estimates of β and σ with β_1 held fixed, and let $\tilde{\varepsilon}_i = (y_i - x_i^{\mathrm{T}}\tilde{\beta})/\tilde{\sigma}$. The first matrix in the numerator of (8.33) has one column of length $p+1$, given by $\varphi(\widehat{\beta}, \widehat{\sigma}) - \varphi(\tilde{\beta}, \tilde{\sigma})$, and p columns of length $p+1$, given by $\varphi_\lambda(\tilde{\beta}, \tilde{\sigma})$. After some simplification, noting that the score equations defining $\widehat{\beta}$ and $\tilde{\beta}$ are

$$\sum_{i=1}^{n} g'(\tilde{\varepsilon}_i)x_{ij} = 0, \; j = 2, \ldots p, \qquad \sum_{i=1}^{n} g'(\widehat{\varepsilon}_i)x_{ij} = 0, \; j = 1, \ldots p,$$

we have that this matrix is given by

$$\frac{1}{\tilde{\sigma}^2}\begin{pmatrix} \tilde{\sigma}\tilde{g}_1^{\mathrm{T}}X_1 & X_1^{\mathrm{T}}\tilde{G}_2X_{-1} & X_1^{\mathrm{T}}\tilde{G}_2\tilde{\varepsilon} + \tilde{g}_1^{\mathrm{T}}X_1 \\ 0_{p-1} & X_{-1}^{\mathrm{T}}\tilde{G}_2X_{-1} & X_{-1}^{\mathrm{T}}\tilde{G}_2\tilde{\varepsilon} \\ \tilde{\sigma}\tilde{g}_1^{\mathrm{T}}\widehat{\varepsilon} - n\tilde{\sigma}^2/\widehat{\sigma} & \widehat{\varepsilon}^{\mathrm{T}}\tilde{G}_2X_{-1} & \widehat{\varepsilon}^{\mathrm{T}}\tilde{G}_2\tilde{\varepsilon} + \tilde{g}_1^{\mathrm{T}}\widehat{\varepsilon} \end{pmatrix},$$

where X_1 is the first column of the design matrix, X_{-1} is the matrix of the remaining columns, $\tilde{g}_1$ is the vector $\{g'(\tilde{\varepsilon}_1), \ldots, g'(\tilde{\varepsilon}_n)\}^{\mathrm{T}}$, $\tilde{G}_2$ is the diagonal matrix with ith diagonal element $g''(\tilde{\varepsilon}_i)$, and $\widehat{G}_2$ is the diagonal matrix with ith diagonal element $g''(\widehat{\varepsilon}_i)$.

The information matrices needed are

$$j(\widehat{\theta}) = \frac{1}{\widehat{\sigma}^2}\begin{pmatrix} X^{\mathrm{T}}\widehat{G}_2X & X^{\mathrm{T}}\widehat{G}_2\widehat{\varepsilon} \\ \widehat{\varepsilon}^{\mathrm{T}}\widehat{G}_2X & \widehat{\varepsilon}^{\mathrm{T}}\widehat{G}_2\widehat{\varepsilon} + n \end{pmatrix}$$

and

$$j_{\lambda\lambda}(\tilde{\theta}) = \frac{1}{\tilde{\sigma}^2} \begin{pmatrix} X_{-1}^{\mathrm{T}} \tilde{G}_2 X_{-1} & X_{-1}^{\mathrm{T}} \tilde{G}_2 \tilde{\varepsilon} \\ \tilde{\varepsilon}^{\mathrm{T}} \tilde{G}_2 X_{-1} & \tilde{\varepsilon}^{\mathrm{T}} \tilde{G}_2 \tilde{\varepsilon} + n \end{pmatrix}.$$

The expression for q_2 obtained by combining these as indicated by (8.33) is identical to that for q_1 obtained by using the Barndorff-Nielsen version (8.32), because in this model there is an exact ancillary statistic, with components $\hat{\varepsilon}_i = (y_i - x_i^{\mathrm{T}} \hat{\beta})/\hat{\sigma}$, and the vectors making up the matrix V are tangent to that exact ancillary. Thus $\ell_{;\hat{a}} = \ell_{;V}$.

The explicit expression for q_2 obtained by combining these factors is fairly complex, but can be expressed in much simpler form when computed in the full location parametrization $(\beta, \log \sigma)$; in this case

$$q_1 = q_2 = j_{\mathrm{p}}(\hat{\psi})^{-1/2} \ell'_{\mathrm{p}}(\psi) \rho(\psi, \hat{\psi})^{-1}, \tag{8.49}$$

where ψ is any component of β, or $\log \sigma$, and ρ is given by (8.46). This is the score statistic corrected by the factor ρ^{-1}; compare to (8.45). Here we can write $r_{\mathrm{NP}} = -r^{-1} \log \rho$, and treat this component of $r^* - r$ as the adjustment due to eliminating the nuisance parameters.

Because an exact ancillary statistic is available, it is also possible to compute the derivative needed to define the modified profile log likelihood $\ell_{\mathrm{m}}(\psi)$. When the parameter of interest is β_1 this is

$$
\begin{aligned}
\ell_{\mathrm{m}}(\sigma) = {}& -(n-p) \log \tilde{\sigma} - \sum_{i=1}^{n} g(\tilde{\varepsilon}_i) - \frac{1}{2} \log |X_{-1}^{\mathrm{T}} \tilde{G}_2 X_{-1}| \\
& - \log |\hat{\varepsilon}^{\mathrm{T}} \tilde{G}_2 \hat{\varepsilon} + \tilde{g}_1 \hat{\varepsilon} - \hat{\varepsilon}^{\mathrm{T}} \tilde{G}_2 X_{-1} (X_{-1}^{\mathrm{T}} \tilde{G}_2 X_{-1})^{-1} X_{-1}^{\mathrm{T}} \tilde{G}_2 \hat{\varepsilon}| \\
& + \frac{1}{2} \log |\tilde{\varepsilon}^{\mathrm{T}} \tilde{G}_2 \tilde{\varepsilon} + n - \tilde{\varepsilon}^{\mathrm{T}} \tilde{G}_2 X_{-1} (X_{-1}^{\mathrm{T}} \tilde{G}_2 X_{-1})^{-1} X_{-1}^{\mathrm{T}} \tilde{G}_2 \tilde{\varepsilon}|.
\end{aligned}
$$

As in the linear exponential family, a tail area approximation based on ℓ_{m} is given by

$$\Phi(r_{\mathrm{m}}^*) = \Phi \left\{ r_{\mathrm{m}} + \frac{1}{r_{\mathrm{m}}} \log \left(\frac{q_{\mathrm{m}}}{r_{\mathrm{m}}} \right) \right\}, \tag{8.50}$$

where

$$
\begin{aligned}
r_{\mathrm{m}} &= \mathrm{sign}(\hat{\psi}_{\mathrm{m}} - \psi)[2\{\ell_{\mathrm{m}}(\hat{\psi}_{\mathrm{m}}) - \ell_{\mathrm{m}}(\psi)\}]^{1/2}, \\
q_{\mathrm{m}} &= j_{\mathrm{m}}(\hat{\psi}_{\mathrm{m}})^{-1/2} \ell_{m'}(\psi),
\end{aligned}
$$

and this is also accurate to $O(n^{-3/2})$. Limited numerical evidence suggests that (8.50) may be more accurate in finite samples than the approximation based on the profile log likelihood, but it is usually more difficult to compute.

Approximations using $q_1 = q_2$ are the basis of the `marg` package.

8.6.3 Nonlinear regression

In the previous two examples the same expression was obtained for the correction terms q_1 and q_2, that is, it made no difference whether one used the p^* or the p_{TEM} density approximations. This is no longer true if we consider the nonlinear model

$$y_{ij} = \mu(x_i; \beta) + w(x_i; \beta, \rho)\varepsilon_{ij}, \quad i = 1, \ldots, m, \quad j = 1, \ldots, n_i,$$

where ε_{ij} is a standard normal error, and $\mu(x_i; \beta)$ and $w(x_i; \beta, \rho)^2$ are known mean and variance functions indexed by the d-dimensional parameter $\theta = (\beta, \rho)$.

We start with the derivation of the q_2 correction term. In Section 8.4.2 we obtained the expression for the canonical parameter

$$\varphi(\theta)^{\mathrm{T}} = -g_1^{\mathrm{T}} W^{-1} \left(\widehat{X}_\beta + \widehat{\mathcal{E}}\widehat{W}_\beta \quad \widehat{\mathcal{E}}\widehat{W}_\rho \right)$$

of the tangent exponential model. As in the previous example let $\tilde{\beta}$ and $\tilde{\rho}$ denote the constrained maximum likelihood estimates of β and ρ, where the parameter of interest is a scalar component of $\theta = (\beta, \rho)$, and let $\tilde{\varepsilon}_{ij} = \{y_{ij} - \mu(x_i; \tilde{\beta})\}/w(x_i; \tilde{\beta}, \tilde{\rho})$. For calculating the q_2 correction term we need the d-dimensional vector $\varphi(\widehat{\beta}, \widehat{\rho}) - \varphi(\tilde{\beta}, \tilde{\rho})$ and the $d \times (d-1)$ matrix $\varphi_{\lambda^{\mathrm{T}}}(\tilde{\beta}, \tilde{\rho})$. Some matrix algebra yields

$$\varphi(\widehat{\beta}, \widehat{\rho}) = -\left(\begin{matrix} \widehat{W}_\beta^{\mathrm{T}} \\ \widehat{W}_\rho^{\mathrm{T}} \end{matrix} \right) \widehat{W}^{-1} 1_n$$

and

$$\varphi(\tilde{\beta}, \tilde{\rho}) = -\left(\begin{matrix} \widehat{X}_\beta^{\mathrm{T}} + \widehat{W}_\beta^{\mathrm{T}}\widehat{\mathcal{E}} \\ \widehat{W}_\rho^{\mathrm{T}}\widehat{\mathcal{E}} \end{matrix} \right) \tilde{W}^{-1}\tilde{g}_1,$$

where to obtain the first expression we exploited the score equations

$$\widehat{g}_1^{\mathrm{T}}\widehat{W}^{-1}(\widehat{X}_\beta + \widehat{\mathcal{E}}\widehat{W}_\beta) = 1_n^{\mathrm{T}}\widehat{W}^{-1}\widehat{W}_\beta$$

and

$$\widehat{g}_1^{\mathrm{T}}\widehat{W}^{-1}\widehat{\mathcal{E}}\widehat{W}_\rho = 1_n^{\mathrm{T}}\widehat{W}^{-1}\widehat{W}_\rho.$$

We furthermore have

$$\varphi_\theta(\tilde{\beta}, \tilde{\rho}) = \left(\begin{matrix} \widehat{X}_\beta^{\mathrm{T}} + \widehat{W}_\beta^{\mathrm{T}}\widehat{\mathcal{E}} \\ \widehat{W}_\rho^{\mathrm{T}}\widehat{\mathcal{E}} \end{matrix} \right) \left(\tilde{W}^{-2}(\tilde{G}_2\tilde{X}_\beta + \tilde{G}_1\tilde{W}_\beta) \quad \tilde{W}^{-2}(\tilde{G}_2\tilde{\mathcal{E}} + \tilde{G}_1)\tilde{W}_\rho \right),$$

where G_1 and G_2 are $n \times n$ diagonal matrices with diagonal elements $g'(\varepsilon_{ij})$ and $g''(\varepsilon_{ij})$ and a double prime denoting twice differentiation. The matrix φ_λ is obtained by omitting from φ_θ the column corresponding to the parameter of interest.

In the special case of normal errors we have

$$\varphi(\widehat{\beta}, \widehat{\rho}) = -\begin{pmatrix} \widehat{W}_\beta^\mathrm{T} \\ \widehat{W}_\rho^\mathrm{T} \end{pmatrix} \widehat{W}^{-1} 1_n,$$

$$\varphi(\tilde{\beta}; \tilde{\rho}) = -\begin{pmatrix} \widehat{X}_\beta^\mathrm{T} + \widehat{W}_\beta^\mathrm{T}\widehat{\varepsilon} \\ \widehat{W}_\rho^\mathrm{T}\widehat{\varepsilon} \end{pmatrix} \widetilde{W}^{-1}\tilde{\varepsilon} 1_n,$$

$$\varphi_\theta(\tilde{\beta}, \tilde{\rho}) = \begin{pmatrix} \widehat{X}_\beta^\mathrm{T} + \widehat{W}_\beta^\mathrm{T}\widehat{\varepsilon} \\ \widehat{W}_\rho^\mathrm{T}\widehat{\varepsilon} \end{pmatrix} \begin{pmatrix} \widetilde{W}^{-2}(\widetilde{X}_\beta + \widetilde{\varepsilon}\widetilde{W}_\beta) & 2\widetilde{W}^{-2}\tilde{\varepsilon}\widetilde{W}_\rho \end{pmatrix}.$$

If the variance is constant, we are in the classical nonlinear model and

$$\varphi(\widehat{\beta}; \widehat{\rho}) = -\frac{1}{\widehat{\sigma}} \begin{pmatrix} 0_p \\ n \end{pmatrix},$$

$$\varphi(\tilde{\beta}; \tilde{\rho}) = -\frac{1}{\tilde{\sigma}} \begin{pmatrix} \widehat{X}_\beta^\mathrm{T} \\ 1_n^\mathrm{T}\widehat{\varepsilon} \end{pmatrix} \tilde{\varepsilon} 1_n,$$

$$\varphi_\theta(\tilde{\beta}, \tilde{\rho}) = \frac{1}{\widehat{\sigma}^2} \begin{pmatrix} \widehat{X}_\beta^\mathrm{T} \\ 1_n^\mathrm{T}\widehat{\varepsilon} \end{pmatrix} \begin{pmatrix} \widetilde{X}_\beta & 2\tilde{\varepsilon} 1_n \end{pmatrix}.$$

Some other special cases are considered in Problem 62.

The ingredients needed for the q_2 correction term were obtained above for an arbitrary known continuous density $f(\varepsilon)$. For the q_1 correction term we restrict our attention to the normal case. In particular, we will use Skovgaard's approximations (8.37) to the sample space derivatives. This is possible as model (8.17) can be written as a curved exponential family of order $(2m, d)$, where d is the dimension of the parameter $\theta = (\beta, \rho)$ and m is the number of design points. The canonical parameter of the embedding full family is

$$\phi(\theta) = \left(\frac{\mu(x_1; \beta)}{w(x_1; \beta, \rho)^2}, \dots, \frac{\mu(x_m; \beta)}{w(x_m; \beta, \rho)^2}, -\frac{1}{2w(x_1; \beta, \rho)^2}, \dots, -\frac{1}{2w(x_m; \beta, \rho)^2} \right)^\mathrm{T},$$

and the corresponding sufficient statistic is

$$v(y) = \left(\sum_{j=1}^{n_1} y_{1j}, \dots, \sum_{j=1}^{n_m} y_{mj}, \sum_{j=1}^{n_1} y_{1j}^2, \dots, \sum_{j=1}^{n_m} y_{mj}^2 \right)^\mathrm{T}$$

with covariance matrix

$$\Sigma(\theta) = \begin{pmatrix} \mathrm{diag}(n_i w_i^2) & \mathrm{diag}(2n_i w_i^2 \mu_i) \\ \mathrm{diag}(2n_i w_i^2 \mu_i) & \mathrm{diag}(2n_i w_i^4 + 4w_i^2 \mu_i^2) \end{pmatrix},$$

where $\mu_i = \mu(x_i; \beta)$ and $w_i = w(x_i; \beta, \rho)$. We are using the notation ϕ for the canonical parameter of the embedding model, as it is not the same as the canonical parametrization φ of the tangent exponential model. Elsewhere in this book, ϕ denotes the density function of the standard normal distribution.

To compute the q_1 correction term we need the expected information matrix

$$i(\theta) = \frac{\partial \phi(\theta)^\mathrm{T}}{\partial \theta} \Sigma(\theta) \frac{\partial \phi(\theta)}{\partial \theta^\mathrm{T}},$$

where $\partial \phi(\theta) / \partial \theta^\mathrm{T}$ is the Jacobian of ϕ with respect to θ, and

$$S(\widehat{\theta}, \tilde{\theta}) = \left. \frac{\partial \phi(\theta_1)^\mathrm{T}}{\partial \theta_1} \Sigma(\theta_1) \frac{\partial \phi(\theta_2)}{\partial \theta_2^\mathrm{T}} \right|_{\theta_1 = \widehat{\theta}, \theta_2 = \tilde{\theta}},$$

$$Q(\widehat{\theta}, \tilde{\theta}) = \left. \frac{\partial \phi(\theta_1)^\mathrm{T}}{\partial \theta_1} \Sigma(\theta_1) \{\phi(\theta_1) - \phi(\theta_2)\} \right|_{\theta_1 = \widehat{\theta}, \theta_2 = \tilde{\theta}}.$$

The previous expressions can be written in matrix notation as

$$\phi(\theta) = \begin{pmatrix} M \\ -\frac{1}{2} I_m \end{pmatrix} \mathcal{W}^{-2} 1_m,$$

and

$$\Sigma(\theta) = \begin{pmatrix} N\mathcal{W}^2 & 2N\mathcal{W}^2 M \\ 2NM\mathcal{W}^2 & 4M\mathcal{W}^2 M + 2N\mathcal{W}^4 \end{pmatrix},$$

where M, $\mathcal{W}$ and N are $m \times m$ diagonal matrices with elements $\mu(x_i; \beta)$, $w(x_i; \beta, \rho)$ and n_i, X_β and $\mathcal{W}_\beta$ are $m \times p$ matrices with ith row given by the gradients $\partial \mu(x_i; \beta)/\partial \beta^\mathrm{T}$ and $\partial w(x_i; \beta, \rho)/\partial \beta^\mathrm{T}$, and $\mathcal{W}_\rho$ is a $m \times q$ matrix with ith row given by $\partial w(x_i; \beta, \rho)/\partial \rho^\mathrm{T}$. The quantities I_m and 1_m represent respectively the $m \times m$ identity matrix and a m-dimensional vector of ones. Then

$$\frac{\partial \phi(\theta)}{\partial \theta^\mathrm{T}} = \mathcal{W}^{-3} \begin{pmatrix} \mathcal{W}X_\beta - 2M\mathcal{W}_\beta & -2M\mathcal{W}_\rho \\ \mathcal{W}_\beta & \mathcal{W}_\rho \end{pmatrix}$$

and

$$\frac{\partial \phi(\theta)^\mathrm{T}}{\partial \theta} \Sigma(\theta) = \begin{pmatrix} X_\beta^\mathrm{T} N & 2(X_\beta^\mathrm{T} M + \mathcal{W}_\beta^\mathrm{T} \mathcal{W}) N - 4\mathcal{W}_\beta^\mathrm{T} \mathcal{W}^{-1} M^2 (N - I_m) \\ 0_{q \times m} & -4\mathcal{W}_\rho^\mathrm{T} \mathcal{W}^{-1} M^2 (N - I_m) + 2\mathcal{W}_\rho \mathcal{W} N \end{pmatrix},$$

where $0_{q \times m}$ is a $q \times m$ matrix of zeros.

The approximations obtained using the above expressions for the two correction terms q_1 and q_2 are the basis of the R package `nlreg` in the bundle `hoa`.

8.7 Approximations for Bayesian inference

In the Bayesian setting with a prior density $\pi(\theta)$ for θ, the analogue of the first order asymptotic theory summarized in Section 8.3 is the asymptotic normality of the posterior density for θ. Write

$$\pi(\theta \mid y) = \frac{f(y \mid \theta) \pi(\theta)}{\int f(y \mid \theta) \pi(\theta) \mathrm{d}\theta}, \tag{8.51}$$

where the notation $f(y \mid \theta)$ emphasizes the interpretation as a conditional density. For scalar θ, let $a_n = \widehat{\theta} + aj(\widehat{\theta})^{-1/2}$ and $b_n = \widehat{\theta} + bj(\widehat{\theta})^{-1/2}$. Under regularity conditions on the model similar to those required for asymptotic normality of the maximum likelihood estimator, and under conditions on the prior, the posterior is asymptotically normal:

$$\int_{a_n}^{b_n} \pi(\theta \mid y)\mathrm{d}\theta \overset{\mathrm{p}}{\to} \Phi(b) - \Phi(a)$$

as $n \to \infty$, where $\overset{\mathrm{p}}{\to}$ denotes convergence in probability. The first order approximation from this result is that $\theta \mid y \overset{\cdot}{\sim} N\{\widehat{\theta}, j(\widehat{\theta})^{-1}\}$. The analogous argument in the presence of nuisance parameters applies to the marginal posterior for ψ, giving the approximation $\psi \mid y \overset{\cdot}{\sim} N(\widehat{\psi}, j_p(\widehat{\psi})^{-1})$.

A p^*-like approximation to the posterior density for θ can be obtained by the technique of Laplace approximation. Taylor series expansion of the integrand in the denominator of (8.51) around $\widehat{\theta}$ gives

$$\pi(\theta \mid y) \overset{\cdot}{=} \{\sqrt{(2\pi)}\}^{-d} |j(\widehat{\theta})|^{1/2} \exp\{\ell(\theta) - \ell(\widehat{\theta})\}\{\pi(\theta)/\pi(\widehat{\theta})\}, \tag{8.52}$$

which when renormalized is identical in form to the p^* approximation (8.12) for flat priors π, although the interpretation is different.

The derivation of a tail area approximation from the posterior density is simpler than that outlined in Section 8.5, as it involves only integrations over the parameter space. First, in the scalar parameter setting, we start with the Laplace approximation to the density given at (8.52), with $d = 1$. We then have the posterior distribution function

$$\Pi(\theta^0 \mid y) = \int_{-\infty}^{\theta^0} \pi(\theta \mid y)\,\mathrm{d}\theta$$

$$\overset{\cdot}{=} \int_{-\infty}^{r^0} c\phi(r) \left(\frac{r}{q_B}\right) \mathrm{d}r$$

$$\overset{\cdot}{=} \Phi(r^0) + \phi(r^0) \left(\frac{1}{r^0} - \frac{1}{q_B^0}\right),$$

where now

$$q_B^0 = -\ell'(\theta^0)j(\widehat{\theta})^{-1/2}\frac{\pi(\widehat{\theta})}{\pi(\theta^0)}$$

arises from the change of variable from θ to r. Here r is the likelihood root, as in the frequentist case.

When nuisance parameters are present the starting point is the marginal posterior density for ψ,

$$\pi_{\mathrm{m}}(\psi \mid y) = \int \pi(\theta \mid y)\,\mathrm{d}\lambda$$

$$= \int \exp \ell(\theta; y)\pi(\theta)\,\mathrm{d}\lambda \bigg/ \int \exp \ell(\theta; y)\pi(\theta)\,\mathrm{d}\theta$$

$$\overset{\cdot}{=} c|j_p(\widehat{\psi})|^{1/2}\exp\{\ell_p(\psi) - \ell_p(\widehat{\psi})\}\rho(\psi, \widehat{\psi})\frac{\pi(\psi, \widehat{\lambda}_\psi)}{\pi(\widehat{\psi}, \widehat{\lambda})}, \tag{8.53}$$

where ρ is defined at (8.46).

The Laplace approximation (8.53) is derived by using (8.52) for the denominator and the analogous version of (8.52) with ψ held fixed in the numerator. An application of the usual tail area argument now gives an approximation of the form (8.21) or (8.22) with

$$q_{\mathrm{B}} = -\ell_{\mathrm{p}}'(\psi)\, j_{\mathrm{p}}(\widehat{\psi})^{-1/2} \rho(\psi, \widehat{\psi})^{-1} \pi(\widehat{\psi}, \widehat{\lambda}) / \pi(\psi, \widehat{\lambda}_\psi).$$

Illustration: Non-normal linear regression

The model (8.48) has the natural non-informative prior $\pi(\beta, \sigma) \propto d\beta\, d\sigma/\sigma$, and the calculation of q using the above Bayesian approximations is also straightforward. For β_1 the parameter of interest, the result is

$$q_{\mathrm{B}} = -\frac{1}{\tilde\sigma} \tilde g_1^{\mathrm{T}} X_1 \frac{1}{\tilde\sigma^2} \left| \begin{matrix} X_{-1}^{\mathrm{T}} \tilde G_2 X_{-1} & X_{-1}^{\mathrm{T}} \tilde G_2 \tilde\varepsilon \\ \tilde\varepsilon^{\mathrm{T}} \tilde G_2 X_{-1} & \tilde\varepsilon^{\mathrm{T}} \tilde G_2 \tilde\varepsilon + n \end{matrix} \right|^{1/2} |j(\widehat\theta)|^{-1/2},$$

which is equal to the frequentist version given in Section 8.6.2. In the case of normal errors, both approximations give q as a function of the usual t-statistic.

Non-informative priors

If ψ is scalar it is possible to reparametrize the model in terms of ψ and an orthogonal parameter $\eta = \eta(\psi, \lambda)$, chosen so that $i_{\psi\eta}(\psi, \eta) = 0$ for all ψ and η. Then a class of non-informative priors for ψ is given by

$$\pi(\psi, \eta) \propto i_{\psi\psi}(\psi, \eta)^{1/2} g(\eta) \tag{8.54}$$

for an arbitrary function g of η. These priors are sometimes called frequentist matching priors, or simply matching priors, as they have the property that the $(1 - \alpha)$ posterior probability limit for ψ has confidence coefficient $(1 - \alpha) + O(n^{-1})$:

$$\mathrm{Pr}_{\theta|Y}\{\psi \le \psi^{(1-\alpha)}(y)\} = \mathrm{Pr}_{Y|\theta}\{\psi^{(1-\alpha)}(y) \ge \psi\}\{1 + O(n^{-1})\}.$$

In a full exponential family

$$f(u, v; \theta) = \exp\{\psi u + \lambda^{\mathrm{T}} v - c(\psi, \lambda) + h(y)\},$$

the parameter orthogonal to ψ is $\eta(\psi, \lambda) = \mathrm{E}(V; \psi, \lambda)$, where the statistic V associated with λ has observed value v. In this model the maximum likelihood estimate $\widehat\lambda_\psi$ for fixed ψ is determined by the equation

$$v = \mathrm{E}(V; \psi, \widehat\lambda_\psi) = \eta(\psi, \widehat\lambda_\psi) = \widehat\eta_\psi,$$

which shows that $\widehat{\eta}_\psi$ is constant in ψ. Thus in the orthogonal parametrization we have

$$\frac{\pi(\widehat{\psi}, \widehat{\eta})}{\pi(\psi, \widehat{\eta}_\psi)} = \frac{i_{\psi\psi}(\widehat{\psi}, \widehat{\eta})^{1/2} g(\widehat{\eta})}{i_{\psi\psi}(\psi, \widehat{\eta}_\psi)^{1/2} g(\widehat{\eta}_\psi)} = \frac{i_{\psi\psi}(\widehat{\psi}, \widehat{\eta})^{1/2}}{i_{\psi\psi}(\psi, \widehat{\eta})^{1/2}} = \frac{j_{\psi\psi}(\widehat{\psi}, \widehat{\eta})^{1/2}}{j_{\psi\psi}(\psi, \widehat{\eta})^{1/2}}$$

because observed and expected information matrices for such models are equal. Also, since $j_{\lambda\lambda} = j_{\eta\eta}^{-1}$, we have

$$q_{\mathrm{B}} = -\ell_{\mathrm{p}}'(\psi) i_{\psi\psi}(\psi, \widehat{\eta}_\psi)^{-1/2} \frac{|i_{\lambda\lambda}(\widehat{\psi}, \widehat{\lambda})|^{1/2}}{|i_{\lambda\lambda}(\psi, \widehat{\lambda}_\psi)|^{1/2}}, \tag{8.55}$$

where

$$i_{\psi\psi}(\psi, \widehat{\eta}_\psi) = i_{\psi\psi}(\widehat{\theta}_\psi) - i_{\psi\lambda}(\widehat{\theta}_\psi) i_{\lambda\lambda}(\widehat{\theta}_\psi)^{-1} i_{\lambda\psi}(\widehat{\theta}_\psi),$$

so there is no need to explicitly compute with the orthogonalized parameter. This is illustrated in Section 4.2.

The quantities given in this section can also be used to obtain the correction term in (8.43) leading to the adjusted log likelihood function $\ell_{\mathrm{F}}(\psi)$ and to compute Skovgaard's multivariate w^* statistic (8.61), discussed in the next section.

8.8 Vector parameters of interest

8.8.1 Introduction

The Lugannani–Rice approximation (8.21) and the Barndorff-Nielsen r^* approximation (8.22) presuppose that the parameter of interest is scalar. In most problems interest focuses on more than one of the parameters in the model, and we apply these scalar approximations to each component of interest separately. It is sometimes of interest however to construct inference for a vector parameter of interest, and there are two higher order approximations that can be used in this context. We first discuss Bartlett correction of the likelihood ratio statistic, and then Skovgaard's statistic w^*.

8.8.2 Bartlett correction

The likelihood ratio statistic for a vector parameter of interest is defined in (8.9) by $w(\theta) = 2\{\ell(\widehat{\theta}) - \ell(\theta)\}$. In regular models it has asymptotically a chi-squared distribution with d degrees of freedom, where d is the length of θ. The approximation implied by this result is

$$w(\theta) \sim \chi_d^2 \{1 + O_p(n^{-1})\};$$

the two-sided nature of the approximation means there is no $O(n^{-1/2})$ term. It can be shown that

$$\mathrm{E}_\theta\{w(\theta)\} = d\{1 + b(\theta)/n + O(n^{-2})\} \tag{8.56}$$

and hence that

$$w_1(\theta) = w(\theta)/\{1 + b(\theta)/n\}, \tag{8.57}$$

has expected value $d\{1 + O(n^{-2})\}$.

An explicit expansion for $b(\theta)$ is

$$i^{jk} i^{lm} \left(-\frac{1}{4} \kappa_{j,k,l,m} + \frac{1}{2} \kappa_{jl,km} - \frac{1}{4} \kappa_{jk,lm} - \frac{1}{2} \kappa_{jk,l,m} \right)$$

$$+ i^{jk} i^{lm} i^{no} \left(-\frac{1}{2} \kappa_{j,ln} \kappa_{k,mo} + \frac{1}{2} \kappa_{j,k,l} \kappa_{m,no} + \frac{1}{4} \kappa_{j,kl} \kappa_{m,no} \right.$$

$$\left. + \frac{1}{6} \kappa_{j,l,n} \kappa_{k,m,o} + \frac{1}{4} \kappa_{j,k,l} \kappa_{m,n,o} \right) + O(n^{-2}), \tag{8.58}$$

where i^{jk} is the (j, k) element of the inverse expected information matrix and we have used the convention that summation is performed over repeated indices, and

$$\kappa_{j,k,l} = \text{cum}(\ell_j, \ell_k, \ell_l), \quad \kappa_{jk,l} = \text{cov}(\ell_{jk}, \ell_l), \quad \kappa_{jk,lm} = \text{cov}(\ell_{jk}, \ell_{lm}),$$

$$\kappa_{jk,l,m} = \text{cum}(\ell_{jk}, \ell_l, \ell_m), \quad \kappa_{j,k,l,m} = \text{cum}(\ell_j, \ell_k, \ell_l, \ell_m), \tag{8.59}$$

are joint cumulants of $\ell_j = \partial\ell/\partial\theta_j$, $\ell_{jk} = \partial^2\ell/\partial\theta_j\partial\theta_k$ and so forth.

A detailed examination of its cumulant generating function shows that $w(\theta)$ follows asymptotically a non-central chi-squared distribution with d degrees of freedom and non-centrality parameter $b(\theta)/n + O(n^{-2})$, and the results of Appendix A.4 then imply that $w_1(\theta)$ follows a χ_d^2 distribution with relative error $O(n^{-2})$:

$$w_1(\theta) \sim \chi_d^2\{1 + O_p(n^{-2})\}. \tag{8.60}$$

In the case that $\theta = (\psi, \lambda)$, and $w(\psi) = 2\{\ell(\widehat{\psi}, \widehat{\lambda}) - \ell(\psi, \widehat{\lambda}_\psi)\}$, where ψ is of length q, we have the corresponding approximations

$$w(\psi) \sim \chi_q^2\{1 + O_p(n^{-1})\},$$

$$\text{E}_\theta\{w(\psi)\} = q\{1 + B(\theta)/n + O(n^{-1})\},$$

$$w_1(\psi) \equiv w(\psi)/\{1 + B(\theta)/n\} \sim \chi_q^2\{1 + O(n^{-2})\}.$$

As we can write

$$w(\psi) = 2\{\ell(\widehat{\psi}, \widehat{\lambda}) - \ell(\psi, \lambda)\} - 2\{\ell(\psi, \widehat{\lambda}_\psi) - \ell(\psi, \lambda)\},$$

$B(\theta)$ can be obtained by considering the results for the model with no nuisance parameters.

To compute $w_1(\psi)$ it is necessary to estimate $B(\theta)$ or to estimate the terms in the expansion (8.58), which also depend on θ. It has been shown that estimating $B(\theta)$, by, for example $B(\widehat{\theta})$, does not change the asymptotic properties (8.57) and (8.60). In some models $B(\theta)$ can be computed exactly, and in others explicit expressions for the cumulants in (8.59) will be computable. In general, however, computation of the Bartlett correction is rather complex. The form of (8.56) suggests using the jackknife or bootstrap to estimate $E\{w(\theta)\}$; this is used in Section 7.5.

There is also a Bayesian version of Bartlett correction; see the bibliographic notes and Problem 66.

8.8.3 Skovgaard's approximation

A detailed study of the method of construction of r^* in the scalar parameter case led to the following suggestion by Skovgaard (2001) for an r^*-type statistic in the multiparameter case. Write

$$r^* = r - \frac{1}{r} \log \gamma,$$

where $\gamma = r/q_1$. The approximation to q_1 given at (8.40) entails a corresponding approximation to γ, and an analogous version can be developed from this in the multiparameter case.

The multiparameter version of r^* is

$$w^* = w \left(1 - \frac{1}{w} \log \gamma \right)^2, \tag{8.61}$$

where $w = w(\psi)$, and

$$\gamma = \frac{|i(\widehat{\theta}_\psi)|^{1/2} |i(\widehat{\theta})|^{1/2}}{|\widehat{S}| |j_{\lambda\lambda}(\widehat{\theta}_\psi)^{-1/2}|} \left| [i(\widehat{\theta}_\psi)\widehat{S}^{-1} j(\widehat{\theta}) i(\widehat{\theta})^{-1}\widehat{S}]_{\lambda\lambda} \right|^{-1/2}$$

$$\times \frac{\{\ell^{\mathrm{T}}_{\theta;\widehat{\theta}}(\widehat{\theta}_\psi)\widehat{S}^{-1} i(\widehat{\theta}) j(\widehat{\theta})^{-1}\widehat{S} i(\widehat{\theta}_\psi)^{-1} \ell_{\theta;\widehat{\theta}}(\widehat{\theta}_\psi)\}^{q/2}}{w^{q/2-1} \ell^{\mathrm{T}}_{\theta;\widehat{\theta}}(\widehat{\theta}_\psi)\widehat{S}^{-1}\widehat{Q}}; \tag{8.62}$$

S and Q are as defined at (8.38) and (8.39).

The adjusted profile log likelihood that corresponds to Skovgaard's approximation is

$$\ell_{\mathrm{S}}(\psi) = \ell_{\mathrm{p}}(\psi) + \frac{1}{2} \log |j_{\lambda\lambda}(\psi, \widehat{\lambda}_\psi)| - \log |\widehat{S}_{\lambda\lambda}|.$$

Skovgaard's approximation is illustrated in Sections 5.4 and 6.4, and is implemented in the `nlreg` package.

Bibliographic notes

The notation used in Section 8.2 largely follows that of Barndorff-Nielsen and Cox (1994). The parametrization of a statistical model, usually taken for granted, is rather more subtle than it first appears. A rigorous approach to this is given in McCullagh (2002). Barndorff-Nielsen and Cox (1994, Chapter 2) provides a more thorough discussion of exponential family and transformation family models; there are also book-length treatments of exponential family models (Barndorff-Nielsen, 1978; Brown, 1986).

Numerous books treat the limiting distributions of likelihood based statistics; the summary in Section 8.3 is closest to Chapter 9 of Cox and Hinkley (1974) and Chapter 2 of Barndorff-Nielsen and Cox (1994). One approach to improved inference is adjustment of maximum likelihood estimators and related quantities based on direct computation of mean and variance expansions for them (Cox and Snell, 1968; Shenton and Bowman, 1977); Firth (1993) takes the indirect route of adjusting the score statistic and shows connections with Bayesian inference. The literature on asymptotic theory for likelihood based inference in non-regular models is rather dispersed: Self and Liang (1987) discuss one aspect of the problem, but see also the references in Barndorff-Nielsen and Cox (1994).

The connection between ancillarity and asymptotic theory was first discussed in Efron and Hinkley (1978), who in particular showed that observed Fisher information gives a more appropriate estimate of the variance of the maximum likelihood estimator than does expected Fisher information. Barndorff-Nielsen and Cox (1979) inspired much of the later development of higher order asymptotics with their discussion of statistical applications of Edgeworth and saddlepoint approximations.

The p^* approximation was first developed for general settings in a series of papers appearing in *Biometrika* in 1980 (Barndorff-Nielsen, 1980; Cox, 1980; Hinkley, 1980; Durbin, 1980a,b), although it was pre-figured in Daniels (1958). It was presented in a more general context in Barndorff-Nielsen (1983), and a review of then current developments is given in Reid (1988). There are now a number of books and review papers surveying this material and later research. Barndorff-Nielsen and Cox (1994) remains the definitive reference, but Pace and Salvan (1997) and Severini (2000a) give somewhat more accessible accounts. Skovgaard (1985) outlined the steps needed for a general proof of the p^* approximation, and gave a more complete derivation in Skovgaard (1990). The latter paper explains the role of the approximate ancillarity of a; see also Severini (2000a, Chapter 6). The tangent exponential model was introduced in Fraser (1990), and the use of pivotal statistics to derive ancillary directions in Fraser and Reid (1993, 1995, 2001).

The tail area approximation $\Phi(r^*)$ was derived from the p^* density in Barndorff-Nielsen (1986) and somewhat more accessibly in Barndorff-Nielsen (1990); the main results are summarized in Barndorff-Nielsen and Cox (1994, Chapter 6). The tail area approximation based on the tangent exponential model was introduced in Fraser (1990, 1991) and a number of applications of the methods given in Fraser et al. (1999a).

Tail area approximations for linear exponential families were considered in detail in Fraser et al. (1991), where the distinction was made between the version based on the profile log likelihood and the version based on an approximation to the conditional log likelihood, sometimes called the 'double saddlepoint' and the 'sequential saddlepoint', respectively. DiCiccio *et al.* (1990) gave detailed formulae for location-scale versions, along with several examples. DiCiccio and Martin (1991, 1993) considered the connection between Bayesian and location-scale models in more detail; the latter paper also includes examples comparing the behaviour of $\Phi(r^*)$ with the mean- and variance-corrected version of r mentioned at (8.29). The literature on tail area approximations up to 1995 is reviewed in Reid (1996). The derivation of the explicit expression in Section 8.6.2 for regression-scale models is taken from Brazzale (2000, Chapter 3). The expressions for r and q_1 in terms of the t-statistic in normal theory linear regression are studied in Sartori (2003b) and in work as yet unpublished by Sigfrido Iglesias-Gonzalez.

Skovgaard (1996) suggested the moment-based approximation described in Section 8.5.3; a similar approximation involving estimated moments was described in Severini (1999). Skovgaard (2001) introduces the vector parameter version w^* in the same spirit, and gives a very accessible review of the p^* and r^* approximations. The review paper of Reid (2003) covers similar material with more emphasis on the tangent exponential model.

Little is known about higher order asymptotics in non-regular models. One exception is Castillo and López-Ratera (2006), who develop p^* and r^* approximations for scalar-parameter exponential family models when the true value of the parameter is on the boundary of the parameter space, and apply it to a reliability problem.

Barndorff-Nielsen (1986) introduced the modified profile likelihood based on the derivation of r^*, and Fraser (2003) used the tangent exponential model to develop the adjusted likelihood (8.43). Sartori (2003a) compares profile and modified profile log likelihoods in models with increasing numbers of nuisance parameters, and shows that although first order asymptotic theory is not valid for either likelihood in this setting, the asymptotic behaviour of modified profile log likelihood is more 'resistant' to the number of nuisance parameters. Several examples are discussed in Bellio and Sartori (2006). Ferrari *et al.* (2005) consider modified profile log likelihoods for vector parameters of interest, building on work of Stern (1997).

Precise results on higher order approximations for discrete distributions are difficult to obtain. Pierce and Peters (1992) gave the first careful discussion; see also Severini (2000b) and Butler (2007). It seems likely that the continuity corrected version of r^* gives an approximation at the jump points accurate to $O(n^{-3/2})$, and that the smoothed versions discussed in Davison and Wang (2002) and Davison *et al.* (2006) are accurate to $O(n^{-1})$ at the jump points, but may also be accurate to $O(n^{-1})$ in moderate deviation regions, under suitable regularity conditions.

The derivation of posterior asymptotic normality is outlined in Chapter 4 of Berger (1985). The relationship between p^* and Laplace approximation to the posterior density was given in Davison (1986). Laplace approximation to marginal posterior densities was discussed by Tierney and Kadane (1986). Matching priors of the form (8.54) were derived in Peers (1965) and Tibshirani (1989). These are exploited by DiCiccio and Martin (1993) in both frequentist and Bayesian contexts. Sweeting (2005) extends the matching prior argument to develop a type of local matching prior; this is developed further by Ana-Maria Staicu in work as yet unpublished.

Bartlett (1937) introduced the correction to the likelihood ratio statistic in the problem of comparing several variances. A general formula for Bartlett correction and the resulting distribution of the adjusted likelihood ratio statistic was obtained in a heroic series of calculations by Lawley (1956), whose derivation is explained and simplified in McCullagh (1987, Chapter 6). General expressions for the Bartlett factor are given in DiCiccio and Stern (1993), McCullagh and Cox (1986), Skovgaard (2001), and of course Lawley (1956). G. M. Cordeiro has established explicit results for Bartlett correction factors in a series of single-authored and joint publications; the complete list is at the time of writing best accessed through his web page www.ufrpe.br/gauss. In particular Cordeiro (1987) was the first paper to provide explicit expressions for generalized linear models; for a practical application of this see McCullagh and Nelder (1989, Chapter 15). A general review of Bartlett correction is given in Cordeiro and Cribari-Neto (1996). Bayesian Bartlett correction is discussed in Bickel and Ghosh (1990).

9

Numerical implementation

9.1 Introduction

A superficially appealing way to implement higher order inference procedures would be to write general routines in a computer algebra system such as *Maple* or *Mathematica*, designed so that the user need provide the minimum input specific to his or her problem. One would use these routines to derive symbolic expressions for quantities such as r^*, and then evaluate these expressions numerically, concealing the hideous details in the computer. Many models have special structure which this approach does not exploit, however, leading to burdensome derivations of intermediate quantities which then simplify enormously, and symbolic computation packages are generally ill-suited for numerical work on the scale needed for applied statistics. Thus although computer algebra systems can be powerful tools for research in higher order asymptotics, those currently available are unsuitable for passing on the fruits of that research to potential users. It is possible to interface separate packages for symbolic and numerical computation, but this half-way house is system-dependent and demands knowledge of advanced features of the packages.

A more practicable approach recognises that many classes of models can be treated in a modular way, so higher order quantities can be expressed using a few elementary building-blocks. In some cases these must be computed specifically for the problem at hand, but the rudimentary symbolic manipulation facilities of environments for numerical computing such as R can then be exploited. A technique that we call pivot profiling can then be used to obtain higher order quantities for the range of interest, by computing them over a fixed grid of values between which they are interpolated. This use of building-blocks and of pivot profiling is at the core of the higher order fitting routines of the R package bundle hoa, which implements many of the solutions presented in earlier chapters for logistic and log-linear models, linear non-normal models, and nonlinear heteroscedastic regression models. The hoa bundle also includes a package for conditional simulation.

In the following sections we will illustrate this by referring to the hoa code, but it will be evident that these ideas could be applied more broadly.

9.2 Building-blocks

The key to efficient numerical implementation of higher order asymptotics is to identify the building-blocks into which the required expressions can be decomposed and which are provided or can efficiently be handled by the computing device. In linear exponential families, for instance, routines such as the R function glm suffice to calculate the approximations described in Section 8.6.1. For a given value of the interest parameter ψ, all output needed can be retrieved from two fits of the model, one for joint estimation of ψ and the nuisance parameter λ, and the other for estimation of λ when ψ is fixed. To appreciate this, recall that the modified likelihood root is

$$r^* = r + \frac{1}{r} \log\left(\frac{q}{r}\right),$$

where

$$r = \text{sign}(q)[2\{\ell_p(\widehat{\psi}) - \ell_p(\psi)\}]^{1/2},$$

and for a linear exponential family,

$$q = (\widehat{\psi} - \psi) j_p(\widehat{\psi})^{1/2} \left\{ |j_{\lambda\lambda}(\widehat{\psi}, \widehat{\lambda})| / |j_{\lambda\lambda}(\psi, \widehat{\lambda}_\psi)| \right\}^{1/2}.$$

The elements we need are the maximum likelihood estimates of the parameters $(\widehat{\psi}, \widehat{\lambda})$ and the constrained estimate $\widehat{\lambda}_\psi$, the profile log likelihood $\ell_p(\psi) = \ell(\psi, \widehat{\lambda}_\psi)$ and the corresponding observed Fisher information function $j_p(\psi) = -\ell_p''(\psi)$, the likelihood root and Wald pivots, and the determinant of the observed Fisher information function $j_{\lambda\lambda}(\psi, \lambda)$ evaluated at both $(\widehat{\psi}, \widehat{\lambda})$ and at $(\psi, \widehat{\lambda}_\psi)$. It is straightforward to show that the approximations of Section 8.6.1 can be obtained from the two model fits.

The above idea is not confined to linear exponential families. Section 8.6.2 lists all the likelihood quantities needed to implement higher order inference for linear regression models with non-normal errors. Suppose that the parameter of interest is β_j and that the scale parameter σ is unknown. The expression for the (λ, λ) sub-block of the observed information matrix evaluated at the maximum likelihood estimate, for instance, simplifies to

$$j_{\lambda\lambda}(\widehat{\beta}, \widehat{\sigma}) = \frac{1}{\widehat{\sigma}^2} \begin{pmatrix} X_{-j}^{\mathsf{T}} \widehat{G}_2 X_{-j} & X_{-j}^{\mathsf{T}} \widehat{G}_2 \widehat{\varepsilon} \\ \widehat{\varepsilon}^{\mathsf{T}} \widehat{G}_2 X_{-j} & \widehat{\varepsilon}^{\mathsf{T}} \widehat{G}_2 \widehat{\varepsilon} + n \end{pmatrix},$$

where n is the sample size, $\widehat{\varepsilon}$ the usual ancillary statistic, X_{-j} the model matrix with the jth column deleted, and $\widehat{G}_2 = \text{diag}\{g''(\widehat{\varepsilon}_i)\}$ is an $n \times n$ diagonal matrix whose ith element is given by the second derivative of minus the log density g, evaluated at $\widehat{\varepsilon}_i$. More generally, all one needs to implement approximate conditional inference for regression-scale models are the model matrix X, the standardized residuals $\widehat{\varepsilon}$ and the function g, together with its first two derivatives evaluated at the overall and constrained maximum likelihood estimates of the parameters. These building-blocks are then combined to yield the quantities listed in Section 8.6.2.

The same idea also applies to the higher order solutions for nonlinear heteroscedastic regression models discussed in Sections 8.4 and 8.6.3. Let $\mu(x_i; \beta)$ and $w(x_i; \beta, \rho)^2$ denote the mean and variance functions, respectively. Some algebra establishes that the likelihood quantities involved can be split into blocks – vectors, matrices and multi-way arrays – that only depend upon the overall and constrained maximum likelihood estimates, the functions $\mu(\cdot)$ and $w(\cdot)^2$, and their first two derivatives. Consider, for example, the log likelihood contribution from the ith observation,

$$\ell_i(\theta) = -\frac{1}{2} \log w(x_i; \beta, \rho)^2 - \frac{\{y_i - \mu(x_i; \beta)\}^2}{2w(x_i; \beta, \rho)^2}, \tag{9.1}$$

where $\theta = (\beta, \rho)$ is the d-dimensional parameter. Let μ_{ir} and w_{ir} denote the derivatives of the mean and variance functions with respect to a component θ_r of θ. The second derivative of (9.1) is

$$\ell_{irs} = -\frac{w_{irs}^2 w_i^2 + H_{irs} w_i^2 - w_{irs}^2 H_i}{2w_i^4} + \frac{w_{ir}^2 w_{is}^2}{2w_i^4} + \frac{w_{is}^2 H_{ir}}{2w_i^4} + \frac{w_{ir}^2 H_{is}}{2w_i^4} - \frac{H_i w_{ir}^2 w_{is}^2}{w_i^6}, \tag{9.2}$$

where $H_i = (y_i - \mu_i)^2$, $H_{ir} = -2\mu_{ir}(y_i - \mu_i)$ and $H_{irs} = -2\mu_{irs}(y_i - \mu_i) + 2\mu_{ir}\mu_{is}$. Expression (9.2) represents the basic structure of the observed information function $j = [-\sum_i \ell_{irs}]$, independently of the particular form of the mean and variance functions; here we use $[\cdot]$ to indicate the elements of an array. To calculate the observed information function we exploit R's capacity to do vectorized calculations, which operate on entire vectors, matrices, and arrays. Suppose that the n-dimensional vectors $\mathtt{W} = [w_i^2]$ and $\mathtt{H} = [H_i]$, the $n \times d$ matrices $\mathtt{Wr} = [w_{ir}^2]$ and $\mathtt{Hr} = [H_{ir}]$, and the $n \times d \times d$ arrays $\mathtt{Wrs} = [w_{irs}^2]$ and $\mathtt{Hrs} = [H_{irs}]$ are available. We will see in Section 9.4.3 how to obtain them in R. Then we can calculate the observed information matrix as

```
(1/2) * ( apply(1/W^2*(W*Wrs+W*Hrs-H*Wrs), c(1,2), sum) -
         crossprod(1/W^2*Wr,Wr) - crossprod(1/W^2*Hr, Wr) -
         crossprod(1/W^2*Wr,Hr) ) + crossprod(H/W^3*Wr, Wr)
```

Similar expressions hold for all likelihood quantities appearing in the correction terms.

Illustration: Crying babies data

For the sake of illustration we will compute r^* using the crying babies data (Cox, 1970, page 61), matched pairs of binary observations concerning the crying of babies. The babies were observed on eighteen days and on each day one child was lulled. Interest focuses on whether lulling has any effect on crying. The binary logistic model used for this data postulates that on day j the probability that a treated individual will respond positively, that is will not cry, is given by $\exp(\lambda_j + \psi)/\{1 + \exp(\lambda_j + \psi)\}$ and the corresponding probability for an untreated individual is $\exp(\lambda_j)/\{1 + \exp(\lambda_j)\}$, where ψ represents the treatment effect and the parameters $\lambda_1, \ldots, \lambda_{18}$ are regarded as nuisance parameters accounting for the day effect. The R data frame `babies` in the `hoa` package bundle contains the data set. The variables lull, day, r1 and r2 represent respectively an index vector for treatment, an 18-level factor with one level for each day, and the numbers of 'crying' and 'not crying' babies. Logistic regression models are generalized linear models with canonical link function and binomial responses, so we can fit the model using the glm fitting routine. The output is saved in an object of class glm. Method functions such as summary, coef, resid or plot can be used to extract the desired information.

To obtain the modified likelihood root r^* for testing the hypothesis that $\psi = 1$, we have to fit the model twice, first to provide the full maximum likelihood estimates $(\widehat{\psi}, \widehat{\lambda})$, and then to calculate the constrained maximum likelihood estimate $\widehat{\lambda}_\psi$ by including lull as an offset variable in the model formula, i.e. treating as fixed the model term ψu in the log likelihood based on (8.44):

```
> babies.1 <- glm( formula = cbind(r1, r2) ~ day + lull - 1,
+                  family = binomial, data = babies )
> lull.num <- as.numeric( babies$lull ) - 1
> babies.0 <- update( babies.1, formula = . ~ . - lull +
+                                  offset(lull.num) )
```

All the elements needed to calculate r^* are contained in the two glm objects babies.1 and babies.0. The likelihood ratio statistic $2\{\ell(\widehat{\psi}, \widehat{\lambda}) - \ell(\psi, \widehat{\lambda}_\psi)\}$ equals the difference between the deviances of the two models:

```
> lr <- deviance( babies.0 ) - deviance( babies.1 )
> r <- sign( coef(babies.1)["lullyes"] ) * sqrt( lr )
```

The observed profile information equals the inverse of the squared standard error of $\widehat{\psi}$, and may be obtained from the summary object of babies.1, thus yielding the standardized maximum likelihood estimate:

```
> summary.1 <- summary( babies.1 )
> j.p <- 1 / ( coef(summary.1)["lullyes", "Std. Error"] )^2
> q <- ( coef(summary.1)["lullyes", "Estimate"] - 1 ) /
> +    coef(summary.1)["lullyes", "Std. Error"]
```

The determinant $|j_{\lambda\lambda}(\widehat{\psi}, \widehat{\lambda})|$ is given by the equation $|j(\widehat{\psi}, \widehat{\lambda})| = |j_p(\widehat{\psi})||j_{\lambda\lambda}(\widehat{\psi}, \widehat{\lambda})|$, where $j(\widehat{\psi}, \widehat{\lambda})$ is the inverse of the variance–covariance matrix of $(\widehat{\psi}, \widehat{\lambda})$ and is stored in the summary object of babies.1:

```
> j <- 1 / det( summary.1$cov.unscaled )
> j.1 <- j/j.p
```

The matrix $j_{\lambda\lambda}(\psi, \widehat{\lambda}_\psi)$ has as inverse the variance–covariance matrix of the constrained maximum likelihood estimate, which can be obtained from the summary object of babies.0:

```
> summary.0 <- summary( babies.0 )
> j.0 <- 1 / det( summary.0$cov.unscaled )
```

The resulting value for r^* is obtained by

```
> rho <- sqrt( j.1 / j.0 )
> r + log( rho * q/r ) / r
```

yielding $r^*(1) = 0.337$.

To compute, for example, $r^*(0.5)$ we use the offset 0.5*lull.num; in the next section we describe the computation of r^* as a function of ψ.

9.3 Pivot profiling

Standard R routines for model fitting such as glm use only first order results. The Wald statistic is commonly used for testing and to derive confidence intervals, as it is linear in the parameter of interest, and to calculate it we only need the maximum likelihood estimate and its standard error. An alternative is the likelihood root statistic $r(\psi)$, but though the resulting P-values and confidence intervals are often more accurate, $r(\psi)$ requires the re-fitting of the model under the null hypothesis. Thus a hypothesis test involves two fits, while construction of a confidence interval may involve many more.

The higher order inference routines in the hoa packages calculate all statistics that are not linear in the parameter of interest exactly on a grid of values of ψ, repeatedly fitting the model with different offsets, retrieving the necessary output and computing the required statistics for each fit. Numerical interpolation is then used for intermediate values of ψ, thus enabling construction of the pivots introduced in Chapter 2. This represents the bulk of all higher order inference routines. For instance, the following lines of pseudo-R code iteratively update a logistic/log-linear model and retrieve the quantities required to implement higher order asymptotics.

```
for( i in seq( along = offsetCoef ) )
{
    newFormula <- as.formula( paste( deparse( formula(oldFit) ),
                    "-", offsetName,
                    "+ offset(", offsetCoef[i],
                    "* offsetVal)", collapse = "") )
    newFit$formula <- newFormula
    newFit <- eval( newFit )
    summaryFit <- summary( newFit )
    glmCDev[i] <- deviance( newFit )
    glmCDet[i] <- det( summaryFit$cov.unscaled )
}
```

As we saw in Section 9.2, this amounts to calculating the deviance of the reduced model and $|j_{\lambda\lambda}(\psi, \widehat{\lambda}_\psi)|^{-1}$, the determinant of the covariance matrix of the constrained maximum likelihood estimates. These are then stored in the vectors glmCDev and glmCDet to be used in the subsequent calculation of the first order and higher order statistics illustrated in

Sections 2.3 and 8.6.1. These are saved in an object of class `cond`. By default the required quantities are computed exactly at 20 values of ψ equally spaced over the interval with limits $\widehat{\psi} \pm 3.5$ se$(\widehat{\psi})$ (`offsetCoef` in the pseudo-R code), where se$(\widehat{\psi}) = j_p(\widehat{\psi})^{-1/2}$. Both the number of points and the length of the interval can be changed using arguments to the higher order inference routine `cond`.

Figure 9.1 illustrates this using the calculation of r^* over the interval $(-1, 4)$ for the crying babies data. The bullets represent values that have been calculated exactly, while the solid line is obtained by spline interpolation. For $\psi = 1$ we find the value $r^*(1) = 0.337$ that was calculated by hand in Section 9.2. The horizontal dashed lines correspond to the 2.5% and 97.5% quantiles of the standard normal distribution and are used to read off a 95% confidence interval of $(0.007, 2.76)$ for the coefficient of `lull`. The profiles of all statistics are obtained by spline interpolation. This works well for analytic functions such as the profile and adjusted profile likelihoods, but quantities involving the term $r^{-1} \log(q/r)$ typically have a singularity at $\psi = \widehat{\psi}$, near which the numerical values calculated by R can be unstable. In order to avoid this we implement a hybrid algorithm that splices in a polynomial for values of ψ in a small interval around $\widehat{\psi}$; by default this interval has limits $\widehat{\psi} \pm 0.6$ se$(\widehat{\psi})$. The higher order fitting routine for linear non-normal models included in the `marg` package works in the same way, except that the results are saved in an object of class `marg`.

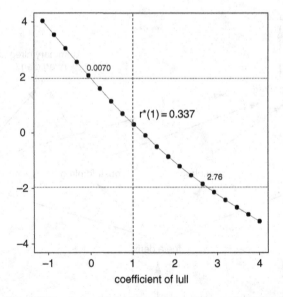

Figure 9.1 Pivot profiling. The core algorithm of the `cond.glm`, `cond.rsm` and `profile.nlreg` higher order method functions of the `hoa` packages enables the computation of confidence intervals by interpolating a spline through a grid of values such as $\{\psi, r^*(\psi)\}$; here the confidence limits are read off as the points at which the fitted spline intersects the desired normal quantiles, shows as horizontal lines.

The higher order methods implemented in the `nlreg` package for nonlinear heteroscedastic models exploit the same key ideas. Estimation of a nonlinear heteroscedastic model is usually more burdensome than is that of a logistic, log-linear, or linear non-normal model. To save execution time, we initially take ψ close to the maximum likelihood estimate and proceed in the two directions $\psi \pm \delta$, where the step size δ is expressed as a fraction of the standard error of $\widehat{\psi}$, $se(\widehat{\psi})$, until the Wald, r and r^* pivots exceed a fixed threshold value, usually 2.2. Thus the number of points at which exact computation is performed may vary, up to a default maximum of 30. In such models there seems to be no need to splice in a polynomial to eliminate the possible singularity at $\psi = \widehat{\psi}$.

Several auxiliary functions have been written to assist the calculation of the higher order solutions using the `profile.nlreg` method. Consider the higher order correction term q given at (8.40), which combines five basic quantities: $\partial\phi(\theta)/\partial\theta$, representing the Jacobian of the canonical parameter ϕ with respect to the parameter $\theta = (\beta, \rho)$, the observed and expected information matrices $j(\theta)$ and $i(\theta)$, and the quantities $S(\theta_1, \theta_2)$ and $Q(\theta_1, \theta_2)$, also given for this model in Section 8.6.3. The corresponding R functions are `theta.deriv`, `obsInfo.nlreg`, `expInfo.nlreg`, `Shat.nlreg` and `qhat.nlreg`. Figure 9.2 shows how they are related.

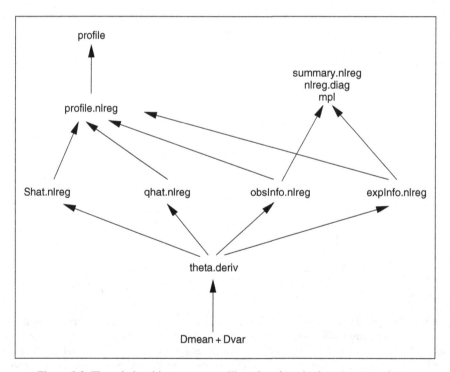

Figure 9.2 The relationships among auxiliary functions in the `nlreg` package.

9.4 Family objects and symbolic differentiation

9.4.1 Introduction

In Section 9.2 we showed that approximate conditional inference for logistic, log-linear, linear non-normal, and nonlinear heteroscedastic regression models requires just a few quantities which characterize the corresponding model class. Some of these quantities, such as formulae for the mean and variance functions, will be provided by the user through the arguments of the fitting routines, and others, such as the model matrix, the constrained and overall maximum likelihood estimates, and the standardized residuals, can easily be calculated in R. The remaining quantities, such as the first and second derivatives of minus the log density of a linear non-normal model or the first two derivatives of the mean and variance functions of a nonlinear regression model, cannot easily be derived during the model fitting phase. Two approaches can be adopted to provide them. The first is to store all quantities in a suitable R object from which they can be retrieved as needed. This works well if the number of models considered is relatively small, as is the case for logistic, log-linear and regression-scale models. For nonlinear heteroscedastic models, however, the number of potential mean and variance functions is nearly unlimited, and it is impracticable to store them all. The second approach uses the `deriv3` routine to calculate the first two derivatives directly in R. We now describe both approaches in more detail.

9.4.2 Family objects

R contains a set of tools to define and handle generalized linear models and hence logistic and log-linear regression models. The foundation on which subsequent computations are built is the class `family`, which saves all the information that specifies a generalized linear model, and in particular its link and variance functions. Figure 9.3 shows how a `family` object is generated. Each model has a generator function of the same name, e.g. `binomial` for a binomial model. These functions invoke a constructor function `make.link`, which retrieves the link function, and create the family object:

```
> binomial( link = logit )
```

```
Family: binomial
Link function: logit
```

```
> binomial( link = logit )$variance
```

```
function (mu)
mu * (1 - mu)
```

Unfortunately, R contains nothing similar that could be used to define a regression-scale model. As we have just seen, the only elements needed in order to deal with non-normal linear regression models are minus the log density g of the corresponding error

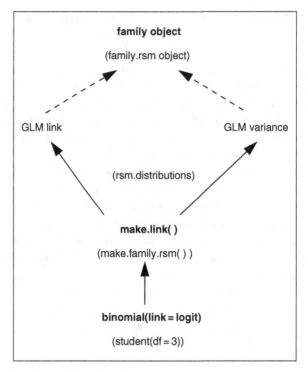

Figure 9.3 How R generates a `family` and a `family.rsm` object.

distribution, and its first two derivatives. Table 9.1 summarizes these for the error densities that are handled by the package `marg`, for the Student *t*, logistic, log-Weibull, extreme-value, Rayleigh, exponential and Huber's least favourable distributions. The exponential and Rayleigh distributions are special cases of the log-Weibull distribution with scale parameter fixed to the values 1 and 0.5, respectively. All these quantities are stored in the data frame `rsm.distributions`. A new class, `family.rsm`, has been introduced. Its design and functioning emulate that for generalized linear models: each error distribution has a generator function, and all invoke the creator function `make.family.rsm`. See Figure 9.3.

```
> student( df = 3 )
student family

g   : function (y, df, ...) (df + 1)/2 * log(1 + y^2/df)
g'  : function (y, df, ...) (df + 1) * y/(df + y^2)
g'' : function (y, df, ...) (df + 1) * (df - y^2)/(df + y^2)^2

df  : 3
```

Table 9.1 *Error distributions handled by the* marg *package.* ν *is the degrees of freedom of the Student* t *distribution. Huber's least favourable density combines a Gaussian centre in the range* $[-k, k]$, *where* k *is the tuning constant, with double exponential tails (Hampel et al., 1986, Figure 1, page 38).*

	Student t	Logistic	log-Weibull	Extreme-value	Huber's least favourable
$g(y)$	$\dfrac{\nu+1}{2}\log\left(1+\dfrac{y^2}{\nu}\right)$	$y+2\log(1+e^{-y})$	e^y-y	$e^{-y}+y$	$\begin{cases} y^2/2, & |y|\le k, \\ k|y|-k^2/2, & |y|>k \end{cases}$
$g'(y)$	$(\nu+1)\dfrac{y}{\nu+y^2}$	$\dfrac{1-e^{-y}}{1+e^{-y}}$	e^y-1	$-e^{-y}+1$	$y\min\left(1,\dfrac{k}{|y|}\right)$
$g''(y)$	$(\nu+1)\dfrac{(\nu-y^2)}{(\nu+y^2)^2}$	$\dfrac{2e^{-y}}{(1+e^{-y})^2}$	e^y	e^{-y}	$I\{|y|\le k\}$
generator	student(df=nu)	logistic()	logWeibull()	extreme()	Huber(k=1.345)

The structure of the `family.rsm` class can easily be reproduced if new models are to be added to the existing ones.

9.4.3 Symbolic differentiation

A nonlinear heteroscedastic model is fully specified by its mean function $\mu(x; \beta)$ and variance function $w(x; \beta, \rho)^2$. In addition, to implement higher order asymptotics we need their first and second derivatives. As mentioned, the R routine `deriv3` can be used to differentiate simple algebraic expressions. It returns a multiple statement expression that, when evaluated, gives the value of the function to be differentiated along with attributes called `gradient` and `hessian`. On request, the returned value is in the form of an R function. The symbolic derivatives are calculated by the R system function D. The principle on which it is based, known as algebraic differentiation, is rather simple. A list of derivatives of elementary functions is stored in the function body as well as the basic rules of the differential operator. A given expression is successively split into sub-expressions until they only contain elementary functions. The chain rule is then applied to obtain the derivative of such a concatenation. The recursive nature of `deriv3` makes it extremely inefficient and slow, so it is hopeless to try and use it to calculate terms such as (8.32) and (8.33) directly. The expressions for the mean and variance functions that are commonly used in nonlinear regression analysis, however, can be handled by `deriv3`. Instead of calculating the higher order correction terms as a whole, we derive the general expression of the derivatives that enter into the building-blocks described on page 172. This is done by the two workhorses of the `nlreg` package – the functions `Dmean` and `Dvar` – which, based upon the function `deriv3`, calculate the first and second derivatives of the mean and the variance functions.

Illustration: Calcium uptake data

To illustrate how the building-blocks are retrieved, we use a simple nonlinear regression model discussed by Davison and Hinkley (1997, Example 7.7). The data concern the calcium uptake of cells y as a function of time x, after being suspended in a solution of radioactive calcium (Rawlings, 1988, page 403); 27 observations in all. The mean function is $\mu(x; \beta) = \beta_0 \{1 - \exp(-\beta_1 x)\}$, β_0 and β_1 are unknown regression coefficients, and the error term $\varepsilon \sim N(0, \sigma^2)$ in (5.3) follows a normal distribution with unknown variance σ^2. Let

```
> calcium.nl <- nlreg( formula = cal ~ b0 * (1 - exp(-b1 * time)),
+                      data = calcium, start = c(b0 = 4.3, b1 = 0.2) )
```

contain the fitted model object. The function md, defined by

```
> md <- Dmean( calcium.nl )
> md
```

```
function (b0, b1, logs)
{
    .expr3 <- exp(-b1 * time)
    .expr4 <- 1 - .expr3
    .expr6 <- .expr3 * time
    .value <- b0 * .expr4
    .grad <- array(0, c(length(.value), 2), list(NULL, c("b0",
        "b1")))
    .hessian <- array(0, c(length(.value), 2, 2), list(NULL,
        c("b0", "b1"), c("b0", "b1")))
    .grad[, "b0"] <- .expr4
    .hessian[, "b0", "b0"] <- 0
    .hessian[, "b0", "b1"] <- .hessian[, "b1", "b0"] <- .expr6
    .grad[, "b1"] <- b0 * .expr6
    .hessian[, "b1", "b1"] <- -(b0 * (.expr6 * time))
    attr(.value, "gradient") <- .grad
    attr(.value, "hessian") <- .hessian
    .value
}
```

returns the mean function, the gradient vector and the Hessian matrix evaluated at the parameter values specified by the arguments b0, b1 and logs, and for each of the 27 data points in the time covariate. The 27×2 matrix of first derivatives can be extracted from the gradient attribute, whereas the hessian attribute contains the $27 \times 2 \times 2$ array of second derivatives.

```
> attach( calcium )
> calcium.md <- md( b0=4.31, b1=0.208, logs=-1.29 )
> names( attributes(calcium.md) )
 [1] "names" "gradient" "hessian"
> attr( calcium.md, "gradient" )
               b0          b1
 [1,]  0.08935305  1.766200
 [2,]  0.08935305  1.766200
 [3,]  0.08935305  1.766200
 [4,]  0.23692580  4.275505
   .         .           .
   .         .           .
   .         .           .
[26,]  0.95584283  2.854761
[27,]  0.95584283  2.854761
```

These matrices correspond to the quantities μ_{ir} and μ_{irs} that appear in the expression for the observed information matrix (9.2). The R objects H, Hr, and Hrs corresponding to H_i, H_{ir} and H_{irs} are easily obtained.

```
grad <- attr( calcium.md, "gradient" )
hess <- attr( calcium.md, "hessian" )
nobs <- nrow( grad ) ; npar <- ncol( grad )
H <- resid( calcium.nl )^2 * calcium.nl$weights
Hr <- - 2 * sqrt(H) * grad
Hrs <- - 2 * sqrt(H) * hess +
+   2 * aperm(array(apply(grad, 1, function(x) crossprod(t(x))),
+                   dim=c(npar, npar, nobs)), perm=c(3, 1, 2))
```

The `Dvar` routine is used to define the variables `W`, `Wr`, and `Wrs` which represent the quantities w_i^2, w_{ir}^2 and w_{irs}^2. The same idea applies to all other likelihood quantities involved in the higher order correction terms.

9.5 Other software

The tangent exponential family described in Section 8.4.2 suggests an approach to higher order inference for general settings, and in this section we briefly describe some software that takes a first step towards this.

Our R function is available from the book's home page and is called `fraser.reid`. It performs the following tasks:

- overall maximization of a log likelihood $\ell(\theta)$, computation of the observed Fisher information matrix $j(\widehat{\theta})$ at the maximum likelihood estimate $\widehat{\theta}$, and of the corresponding standard errors from the diagonal elements of the matrix $j(\widehat{\theta})^{-1}$;
- constrained maximization of $\ell(\theta)$ with respect to the nuisance parameter λ for the required values of the interest parameter ψ, to yield $\widehat{\theta}_\psi = (\psi, \widehat{\lambda}_\psi)$, and computation of the observed Fisher information matrices, $j_{\lambda\lambda}(\widehat{\theta}_\psi)$;
- computation of the local parametrization $\varphi(\theta)$ and its derivatives $\varphi_\theta = \partial\varphi/\partial\theta^\mathrm{T}$ at both $\widehat{\theta}$ and $\widehat{\theta}_\psi$; and finally
- computation of

$$q(\psi) = \frac{\left|\varphi(\widehat{\theta}) - \varphi(\widehat{\theta}_\psi) \quad \varphi_\lambda(\widehat{\theta}_\psi)\right|}{\left|\varphi_\theta(\widehat{\theta})\right|} \left\{ \frac{\left|j(\widehat{\theta})\right|}{\left|j_{\lambda\lambda}(\widehat{\theta}_\psi)\right|} \right\}^{1/2}, \tag{9.3}$$

and profiling of the quantities $r(\psi)$, $r^*(\psi)$, and so forth.

The essential input to `fraser.reid` is a function `nlogL` to compute the negative log likelihood, a function `make.V` to compute and output the required ancillary directions $V_1, \ldots, V_n$, and a function `phi` which uses these and expression (2.12) to find the local parametrization $\varphi(\theta)$.

The `nlogL` function depends on three arguments: the interest parameter ψ, the nuisance parameter λ and the data, contained in `d`. The `phi` function also depends on three arguments: the parameter vector $\theta = (\psi, \lambda)$, the ancillary directions $V_1, \ldots, V_n$ and the data d. The function make.V depends only on the parameter vector θ and on the data.

Summaries 3.4 and 7.1 contain examples of such code. In the first of these the V_i are not needed for the computation of φ, and the function make.V returns a NULL, while in the second the V_i are computed.

Apart from these functions, invocation of fraser.reid requires also the specification of initial values for the overall maximization, specified through th.init. Values of ψ may be specified, as in

```
> fr.test <- fraser.reid( psi = 0, nlogL, phi, make.V,
+                         th.init = initial, d )
```

in which case higher order computations are performed only at $\psi = 0$; this would be required for example for a test of the hypothesis $\psi = 0$. If confidence intervals for ψ are required, then

```
> fr.ci <- fraser.reid( psi = NULL, nlogL, phi, make.V,
+                       th.init = initial, d )
```

will make the computations on a grid of n.psi equally-spaced values of ψ on either side of the maximum likelihood estimate; by default n.psi equals 50. The grid can be specified by setting psi=seq(from=0,to=1,length=11), for example, which performs the computations at $\psi = 0, 0.1, \ldots, 1.0$.

The observed Fisher information and the matrix of derivatives φ_θ needed in (9.3) are computed by crude numerical differencing, using a tolerance tol of value 10^{-5} by default; this can be changed when invoking fraser.reid.

Confidence intervals and graphical output are obtained using the routines lik.ci and plot.fr. By default the first gives equi-tailed 95% confidence intervals based on the Wald pivot, the likelihood root, and the modified likelihood root, obtained by pivot profiling; the level of the interval may be changed by altering the default value of the conf=c(0.975,0.025) argument:

```
> lik.ci( fr.ci, conf = c(0.95, 0.05) )
```

will yield equi-tailed 90% intervals. The command

```
> plot.fr( fr.ci )
```

produces some plots of the object, and.

```
> plot.fr( fr.ci, psi = 0 )
```

produces also P-values for a test of the hypothesis $\psi = 0$.

Potential improvements include the investigation of numerical-analytical aspects, the use of automatic differentiation in place of numerical differencing, and the robustification of the code and output. We leave these as challenges for the sufficiently motivated reader.

Bibliographic notes

The development of tensor calculus for statistical quantities pioneered by McCullagh (1987) paved the way for symbolic calculation of asymptotic expansions in statistics. A short introduction to symbolic computation and its use in statistics is given by Kendall (1998). As outlined there, computer algebra can both assist in the derivation of a general mathematical expression and serve as 'symbolic calculator', performing tasks which would be straightforward if they were smaller in scale. A complete reference on the first topic is the book by Andrews and Stafford (2000), which unifies several results on the study of likelihood and likelihood-type functions. It is accompanied by a set of *Mathematica* notebooks that implement most of the higher order calculations an investigator is likely to face. A system of filters has been developed to extend the general results to specific laws.

Kendall (1992) describes a REDUCE package intended to detect whether an expression that involves many summations over repeated dummy indices is tensorially invariant, and includes functions to manipulate invariant Taylor series expansions. A somewhat less elaborate example is Kabaila (1993), where Edgeworth expansions for a smooth function model are calculated in *Mathematica*. Currie (1995) gives *Mathematica* code for several examples of maximum likelihood estimation, including incomplete data problems. Uusipaikka (1998) has developed *Mathematica* notebooks for likelihood-based inference; see `http://www.statisticalinference.com`. Rose and Smith (2002) describe the results of a similar undertaking. Bellio and Brazzale (2001) provide a REDUCE package for approximate conditional and marginal inference. More details on symbolic calculation in S are given in Becker *et al.* (1988, Section 9.6).

Software implementing higher order inference in numerical computing environments has been made available since the mid-1990s, though usually with a limited range of application. An example is the S-PLUS code that accompanies Field and Welsh (1998), which yields robust confidence intervals for linear regression models. Yi *et al.* (2002) implement Skovgaard's approximation to the modified likelihood root for linear regression and generalized linear models. The first to point out explicitly that the output of standard fitting routines suffices to calculate higher order solutions for linear exponential families was Davison (1988), though the idea is implicit in Barndorff-Nielsen and Cox (1979). It was further developed in in Brazzale (1999) and Bellio and Brazzale (2003), and resulted in the package bundle hoa developed in the R computing language (R Development Core Team, 2005). Both R and hoa are available at CRAN (`http://cran.r-project.org`) or at the book's web site (`http://statwww.epfl.ch/AA`).

10

Problems and further results

In this chapter we give some exercises intended to provide examples of numerical calculation or further theoretical development. We also briefly describe a number of results from the published literature, so the exercises for each chapter serve as an extension to its bibliographic notes. The level of difficulty of the exercises ranges from straightforward application of ideas in the text through to new research.

Chapter 2

1. Assume that $y = (y_1, \ldots, y_n)$ is a vector of independent identically distributed observations from a scalar parameter model $f(y; \theta)$. Show that under suitable smoothness and boundedness conditions on f, an expansion of the score equation $\ell_\theta(\widehat{\theta}; y) = 0$ leads to

$$\sqrt{n}(\widehat{\theta} - \theta) = \frac{1}{\sqrt{n}} \ell_\theta(\theta) / \frac{1}{n} \{-\ell_{\theta\theta}(\theta)\} \{1 + o_p(1)\}. \tag{10.1}$$

Under conditions that ensure that $\ell_\theta(\theta)/\sqrt{n}$ is asymptotically normal with mean 0, and that $\widehat{\theta} \overset{p}{\to} \theta$, argue that (10.1) implies the asymptotic normality of r, s and t defined at (2.1), (2.2) and (2.3).

A careful derivation of these results leads to the conclusion that there exists a consistent and asymptotically normal solution to the score equation. A different approach due to Wald (1949) shows that under some conditions the maximum likelihood estimator is consistent. The distinction arises if the score equation has multiple solutions, or if the likelihood is maximized at a parameter value that is not a root of the score equation. Proving the consistency of the maximum likelihood estimator is more difficult than establishing a central limit theorem for a consistent root of the score equation.

This theory is discussed in many books on theoretical statistics: for example Cox and Hinkley (1974, Chapter 9), Pace and Salvan (1997, Chapter 3) and Severini (2000a, Chapter 4). Lehmann (1986, Chapter 6) is a good introduction to a more rigorous analysis; see also Knight (2000, Chapter 5).

2. The exponential family form of the inverse Gaussian distribution has density

$$f(y; \psi, \lambda) = \frac{\sqrt{\lambda}}{\sqrt{2\pi}} e^{\sqrt{\lambda\psi}} y^{-3/2} \exp\left\{-\frac{1}{2}(\lambda y^{-1} + \psi y)\right\}, \quad y > 0, \lambda > 0, \psi > 0. \quad (10.2)$$

(a) Assume that λ is known, and illustrate the normal approximation to the pivots $r(\psi)$, $t(\psi)$ and $r^*(\psi)$, as in Figure 2.2; it is appropriate to take $q = t$. Use some simulated data to compare these approximations to the exact distribution, using the result that the average $\overline{Y}$ of a random sample has an inverse Gaussian distribution with parameters $(n\lambda, n\psi)$. Investigate the behaviour of the r^* approximation as n increases. The inverse Gaussian distribution is implemented in R in the function `pinvgauss` in the `statmod` package.

(b) For both parameters unknown, pivots for ψ are constructed using the formulae in Section 2.3. Carry out some numerical comparisons in this setting.

(c) If the parameter of interest is $E(Y) = \mu = (\lambda/\psi)^{1/2}$, then (2.11) can be used, with $\theta = (\mu, \lambda)$ and $\varphi(\theta) = (\psi, \lambda)$, where $\psi = \lambda/\mu^2$.

Density approximations for the inverse Gaussian are discussed by Barndorff-Nielsen and Cox (1994, Example 6.2).

3. Show that in the $N(\mu, \sigma^2)$ density, the pivots $r(\mu)$, $s(\mu)$, $t(\mu)$ and $r^*(\mu)$ are all functions of the t-statistic $\sqrt{n}(\overline{y} - \mu)/s$, where $s^2 = (n-1)^{-1} \sum (y_i - \overline{y})^2$.

4. Show that expression (2.11) is invariant to replacement of the $d \times 1$ vector φ by $a + B\varphi$, where the constants a and B are of dimensions $d \times 1$ and $d \times d$, respectively, and B is an invertible matrix.

5. Show that in a full d-dimensional exponential family with the interest parameter ψ a component of the canonical parameter, the general formula for q given at (2.11) simplifies to the expression (2.9).

Chapter 3

6. There are a number of location models in which the r^* approximation illustrated in Section 3.2 performs surprisingly well. Investigate the performance of the approximation by constructing graphs similar to Figure 3.1 for:

(a) the logistic density

$$f(y; \theta) = \frac{\exp(y - \theta)}{\{1 + \exp(y - \theta)\}^2}, \quad y, \theta \in \mathbb{R};$$

(b) the location log-gamma density (with k known)

$$f(y; \theta) = \frac{1}{\Gamma(k)} \exp\{k(y - \theta) - e^{y-\theta}\}, \quad y, \theta \in \mathbb{R}, \quad k > 0.$$

(Fraser, 1990; Barndorff-Nielsen and Chamberlin, 1991)

7. Modify the code in Code 3.2 to compute the maximum relative error of the various approximations to the Poisson cumulative distribution function and to the mid-P function based on r and r^*, with and without continuity correction, over a reasonable range of values of y.

8. Suppose Y has a binomial density with denominator n and probability p:

$$f(y; p) = \binom{n}{y} p^y (1-p)^{n-y}, \quad y = 0, \ldots, n, \quad 0 < p < 1.$$

The Wald pivot leads to two-sided confidence intervals of the form $\widehat{p} \pm z_{\alpha/2} \sqrt{\{\widehat{p}(1 - \widehat{p})/n\}}$, where $\widehat{p} = y/n$.

(a) Plot the coverage of the 95% confidence interval for p, as a function of n, for fixed $p = 0.001$, 0.1 and 0.5 (Brown *et al.*, 2001).

(b) Compare this with the coverage of the intervals based on the normal approximation to the distribution of the likelihood root r and the modified likelihood root r^*.

(c) Investigate the coverage properties when n is fixed and p varies.

Brown *et al.* (2001) show that the posterior interval using the Jeffreys' prior approximates the interval obtained from the mid-P-value using the exact distribution. The development of matching priors discussed in Section 8.7 indicates why this is the case.

9. Calculate the quantities needed for higher order confidence limits for the ratio of probabilities for the astronomer data of Section 3.4. Investigate how they differ from those required for the difference of log odds and the difference of probabilities.

10. In the late 1980s blood plasma products for haemophiliacs were both screened and heat treated, in order to remove any risk of HIV infection. However, a number of cases arose in a Canadian study of HIV infections among haemophiliacs with no other risk factors for HIV. The data in Table 10.1 are extracted from the study of these infections, and show the number of persons infected and non-infected according to the source of plasma product.

(a) Compare the inferences based on the exact P-value, the mid-P-value, and the P-value obtained using the normal approximation to r^*.

(b) The odds ratio for this table is infinite; a solution to this problem often proposed is to add 0.5 to each entry of the table. Motivate this adjustment from a Bayesian argument. Compare inference based on these artificial data with that based on the original data. (Firth, 1993)

11. Confidence intervals for the difference of two binomial probabilities based on the Edgeworth expansion are discussed by Zhou *et al.* (2004). They illustrate their proposed intervals on three data sets from the literature; the corresponding 2×2 tables are given in Table 10.2. Compare the P-value for testing equality of event

Table 10.1 *Source of infection from heat-treated blood products (Neumann et al., 1990).*

	HIV infection	no infection
Received plasma from source A only	6	5
Received plasma from sources A & B	0	5

Table 10.2 *Three 2×2 tables (Zhou et al., 2004). Example 1: positive represents successful diagnosis of prostate cancer; group i is the ith hospital in the study (Tempany et al., 1994). Example 2: events are cot deaths in pairs of twins; group 1 are fraternal twins and group 2 are identical twins (Peterson et al., 1980). Example 3: vaccine trial; group 1 is a new vaccine, group 2 is the old vaccine. Edgeworth CI gives 95% confidence intervals for $p_1 - p_2$, where p_i is probability of an event in column 1 in group i, using an Edgeworth expansion derived in Zhou et al. (2004).*

	Example 1		Example 2		Example 3	
	Positive	Negative	2 events	1 event	Success	Failure
Group 1	18	17	2	8	17	1
Group 2	27	14	1	35	11	7
Edgeworth CI	$(-0.361, 0.074)$		$(-0.024, 0.544)$		$(0.051, 0.497)$	

probabilities using the odds-ratio and the risk difference, based on the approximations of Section 3.4, with that based on the Edgeworth expansion.

12. For two independent samples $(y_{11}, \ldots, y_{1n})$ and $(y_{21}, \ldots, y_{2m})$ from exponential distributions with means μ_1, μ_2, respectively, the parameter $\psi = \mu_1/(\mu_1 + \mu_2)$ is equal to $\Pr(Y_1 < Y_2)$, sometimes called the stress–strength reliability. Compare the exact confidence interval for ψ based on the F distribution with that based on the r^* approximation. Investigate the extension of the problem to accommodate censoring or truncation. (Jiang and Wong, 2007)

13. An alternative expression for q to that given at (2.11) is given in Chapter 8 at (8.34). Using the notation outlined there, show that for the ratio of two exponential means

$$\chi(\widehat{\theta}) = \frac{1}{\widehat{\psi} \, \widehat{\lambda} \sqrt{(\psi^2 + 1)}} (\widehat{\psi} - \psi), \quad \chi(\widehat{\theta}_\psi) = 0,$$

and verify that (8.34) yields (3.10).

14. The exact distribution of $\overline{X}/\overline{Y}$ is used at (3.8) to give an equi-tailed confidence interval. An alternative interval is that determined by the highest density values of

the F distribution. In this approach we define upper and lower confidence limits F_U and F_L by the equations

$$\Pr(F_L \leq F_{2n,2m} \leq F_U) = 1 - 2\alpha,$$

$$f_{2n,2m}(F_L) = f_{2n,2m}(F_U),$$

where $F_{2n,2m}$ is a random variable following the F distribution on $(2n, 2m)$ degrees of freedom and $f_{2n,2m}(x)$ is the corresponding density function. Compare the interval based on this procedure with the approximation in Section 3.5.

15. Find the function q equivalent to (3.10) for the difference of the means of two exponential distributions, corresponding to the log likelihood (3.11).

16. Use the general formula (2.11) to give an expression for q when the parameter of interest is the ratio of two log-normal means, and the difference of two log-normal means.

   The ratio of means of two log-normal distributions is treated in Wu *et al.* (2002); a single log-normal mean is treated in DiCiccio and Martin (1993) and Wu *et al.* (2003). DiCiccio and Martin (1993) also compare the normal approximation to r^* to that of $(r - a)/b$, where a and b^2 are asymptotic approximations to the mean and variance of r. As in the case of Bartlett correction illustrated in Section 7.5, these could be estimated by simulation.

17. A more standard representation for the inverse Gaussian distribution than the exponential family form (10.2) is

$$f(y; \mu, \lambda) = \frac{\lambda}{(2\pi y^3)^{1/2}} \exp\left\{-\frac{\lambda}{2\mu^2 y}(y - \mu)^2\right\}, \quad y > 0, \ \lambda, \mu > 0,$$

where μ is the mean and λ is a scale parameter. The ratio λ/μ is called the shape parameter. Apply this model to the cost data of Section 3.5, with parameter of interest equal to the ratio of the means in the two groups.   (Tian and Wilding, 2005)

18. A. C. M. Wong and J. Wu and their colleagues have illustrated the construction of r^* on a number of single sample and two sample examples. In most cases the approximations are illustrated using simple datasets, as well as via simulation studies. Inference for the *coefficient of variation* parameter σ/μ is discussed for normal, gamma and Weibull models in Wong and Wu (2002). Wu *et al.* (2005) consider inference for the ratio of scale parameters in the Weibull model. Wong and Wu (2001) illustrate inference for a somewhat non-standard parameter in a bivariate normal model.

19. Suppose we have a sample (y_{1i}, y_{2i}), $i = 1, \ldots, n$, from a bivariate normal distribution with means zero, variances 1 and correlation θ.

   (a) Show that this forms a $(2, 1)$ curved exponential family with sufficient statistics $\{\sum(y_{1i}^2 + y_{2i}^2), \sum y_{1i} y_{2i}\}$. It can be embedded in a full exponential family by treating $\alpha_1 \sum(y_{1i}^2 + y_{2i}^2)$ and $\alpha_2 \sum(y_{1i}^2 - y_{2i}^2)$ as independent χ_n^2 random variables, where $\alpha_1 = \alpha/(1+\theta)$ and $\alpha_2 = \alpha/(1-\theta)$.

(b) This embedding enables calculation of an approximately ancillary statistic based on the likelihood root for testing $\alpha = 1$; see Barndorff-Nielsen and Cox (1994, Ex. 7.1).

(c) The approach based on φ and q_2 of (8.33) can also be developed for this example using the pivotal statistics

$$z_{1i} = ((y_{1i} + y_{2i})^2 / \{2(1+\theta)\}, (y_{1i} - y_{2i})^2 / \{2(1-\theta)\}),$$

leading to

$$V_i = ((y_{1i} - \widehat{\theta} y_{2i})/(1 - \widehat{\theta}^2), (y_{2i} - \widehat{\theta} y_{1i})/(1 - \widehat{\theta}^2)),$$
$$\ell_{;V}(\theta) = n\{\theta(t - \widehat{\theta}s) - (s - \widehat{\theta}t)\}/\{(1 - \widehat{\theta}^2)(1 + \widehat{\theta}^2)\},$$

where $s = \sum y_{1i} y_{2i}$ and $t = \sum(y_{1i}^2 + y_{2i}^2)$.   (Reid, 2005)

20. At the time of writing, it is unclear how higher order asymptotic methods are affected by outliers, and whether or not the approximations are more model sensitive than first order approximations, although this is conjectured to be so (Ronchetti and Ventura, 2001). In the cost data of Section 3.5, Group 1 has a single observation (the last) much larger than the others. Investigate the sensitivity of the exact and approximate procedures to this observation.

21. T. J. DiCiccio, G. A. Young and co-workers have suggested that higher order significance levels and confidence intervals for parameters in full exponential families may be obtained by simulation from the constrained maximum likelihood fit. They suggest that the significance level for a given value ψ of the interest parameter be assessed by comparing the observed value $r_{\text{obs}}(\psi)$ of the likelihood pivot with values $r_1^\dagger(\psi), \ldots, r_N^\dagger(\psi)$ of this pivot obtained by fitting the constrained and unconstrained models to N independent copies of the data simulated from the parametric model with parameters $\widehat{\theta}_\psi$, and using the significance level

$$N^{-1} \sum_{i=1}^{N} I\{r_i^\dagger(\psi) \leq r_{\text{obs}}(\psi)\},$$

where $I(\cdot)$ denotes an indicator function.

Different simulations seem needed for each ψ of interest.

Discuss the advantages and disadvantages of this procedure relative to the use of $r^*(\psi)$ with normal approximation, and use it to compute the significance level for testing unit cost ratio for the log-normal model for the cost data.

     (DiCiccio *et al.*, 2001; Lee and Young, 2005)

Table 10.3 *Age (years) and presence/absence (1/0) of coronary heart disease for 100 subjects (Hosmer and Lemeshow, 1989, Chapter 1).*

Age	20	23	24	25	25	26	26	28	28	29	30	30	30	30	30
Disease	0	0	0	0	1	0	0	0	0	0	0	0	0	0	0
Age	30	32	32	33	33	34	34	34	34	34	35	35	36	36	36
Disease	1	0	0	0	0	0	0	1	0	0	0	0	0	1	0
Age	37	37	37	38	38	39	39	40	40	41	41	42	42	42	42
Disease	0	1	0	0	0	0	1	0	1	0	0	0	0	0	1
Age	43	43	43	44	44	44	44	45	45	46	46	47	47	47	48
Disease	0	0	1	0	0	1	1	0	1	0	1	0	0	1	0
Age	48	48	49	49	49	50	50	51	52	52	53	53	54	55	55
Disease	1	1	0	0	1	0	1	0	0	1	1	1	1	0	1
Age	55	56	56	56	57	57	57	57	57	57	58	58	58	59	59
Disease	1	1	1	1	0	0	1	1	1	1	0	1	1	1	1
Age	60	60	61	62	62	63	64	64	65	69					
Disease	0	1	1	1	1	1	0	1	1	1					

Chapter 4

22. Table 10.3 gives data on age and coronary heart disease. Compare 95% confidence intervals for the coefficient of age in a logistic regression model obtained using the cond package with those obtained by approximation to the Bayesian posterior, using the matching prior described in Sections 8.7 and 4.2. This example is also discussed in Sweeting (2005).

23. Suppose in the context of logistic regression with a single covariate that interest focuses on the probability of a response at a fixed value of the covariate: $p_0 = \Pr(Y = 1 \mid x_0)$. Show that inference for $\psi = \text{logit } p_0$ can be carried out using the cond package, with covariate $x - x_0$ and the intercept as an offset. (Yubin *et al.*, 2005)

24. Agresti and Min (2005) perform an extensive assessment of the matching properties of various prior distributions, in the context of comparing two binomial probabilities. Their conclusion is that the frequentist coverage of Bayesian posterior intervals can depend rather strongly on the prior, and that more diffuse priors lead to more accurate frequentist coverage.

25. For the cell phone example show that a continuity-corrected confidence interval (γ_L, γ_U) with level $(1 - \alpha)$ for the parameter γ is obtained by solving the equations

$$\Pr\{\text{Bin}(181, \gamma_L) > 157\} + \frac{1}{2}\Pr\{\text{Bin}(181, \gamma_L) = 157\} \geq \alpha/2,$$

$$\Pr\{\text{Bin}(181, \gamma_U) < 157\} + \frac{1}{2}\Pr\{\text{Bin}(181, \gamma_U) = 157\} \geq \alpha/2,$$

and verify that with $\alpha = 0.05$ the result is $(0.812, 0.911)$; find the corresponding interval for ψ.

In the analysis of the cell phone data the estimate of the relative risk is computed by comparing the 'crash window', the time interval in which the crash occurred, to a 'control window', taken to be the same time window on the previous weekday. This overestimates the risk, because the driver may not have been driving during the control window. A detailed discussion of the choice of the control window is given in Redelmeier and Tibshirani (1997b), and a subsidiary experiment to estimate the driving frequency described in Tibshirani and Redelmeier (1997).

26. Independent pairs of binary observations (R_{0i}, R_{1i}) have success probabilities $\{e^{\lambda_i}/(1+e^{\lambda_i}), e^{\psi+\lambda_i}/(1+e^{\psi+\lambda_i})\}$, for $i = 1, \ldots, n$.

(a) Show that the maximum likelihood estimator of ψ based on a conditional likelihood eliminating the λ_i is $\widehat{\psi}_c = \log(R^{01}/R^{10})$, where R^{01} and R^{10} are respectively the numbers of $(0,1)$ and $(1,0)$ pairs. Does $\widehat{\psi}_c$ tend to ψ as $n \to \infty$?

(b) Write down the unconditional likelihood for ψ and λ, and show that the likelihood equations are equivalent to

$$r_{0i} + r_{1i} = \frac{e^{\widehat{\lambda}_i}}{1+e^{\widehat{\lambda}_i}} + \frac{e^{\widehat{\lambda}_i+\widehat{\psi}}}{1+e^{\widehat{\lambda}_i+\widehat{\psi}}}, \quad i = 1, \ldots, n, \qquad (10.3)$$

$$\sum_{i=1}^{n} r_{1i} = \sum_{i=1}^{n} \frac{e^{\widehat{\lambda}_i+\widehat{\psi}}}{1+e^{\widehat{\lambda}_i+\widehat{\psi}}}.$$

(i) Show that the maximum likelihood estimator of λ_i is ∞ if $r_{0i} = r_{1i} = 1$ and $-\infty$ if $r_{0i} = r_{1i} = 0$; such pairs are not informative. (ii) Use (10.3) to show that $\widehat{\lambda}_i = -\widehat{\psi}/2$ for those pairs for which $r_{0i} + r_{1i} = 1$. (iii) Hence deduce that the unconditional maximum likelihood estimator of ψ is $\widehat{\psi}_u = 2\log(R^{01}/R^{10})$. What is the implication for unconditional estimation of ψ?

27. Derive the elements necessary to apply (2.11) to the risk difference of the myeloma data, given that $\widehat{\psi} = 0.08$.

28. Table 10.4 gives data from Example H of Cox and Snell (1981) on faults in a manufacturing process. Each of 22 batches of raw material was divided into two parts and processed by one of two methods ('standard' and 'modified'). The response is the presence or absence of faults in the finished product. A single covariate, purity index, was measured before the processing. Use the package `cond` to compare first order and higher order inferences for the effects of process and purity.

29. One way to express a Poisson model with so-called 'zero-inflation' is to assume that the density takes the form

$$f(y; \mu, \omega) = \begin{cases} \omega + (1-\omega)e^{-\mu}, & y = 0, \\ (1-\omega)\mu^y e^{-\mu}/y!, & y > 0, \end{cases} \quad \mu > 0, 0 \leq \omega < 1.$$

Table 10.4 *Binary data with two covariates (Cox and Snell, 1981, Example H).*
Response 1 indicates a fault in the finished product and response 0 indicates no fault.

Standard process	0	1	1	0	1	1	0	1	0	1	0
Modified process	0	0	0	1	0	0	0	1	0	0	0
Purity index	7.2	6.3	8.5	7.1	8.2	4.6	8.5	6.9	8.0	8.0	9.1
Standard process	0	1	1	0	1	0	1	0	1	0	0
Modified process	1	1	0	1	0	1	0	0	0	1	0
Purity index	6.5	4.9	5.3	7.1	8.4	8.5	6.6	9.1	7.1	7.5	8.3

Table 10.5 *Dental status in each of eight categories from a prospective survey of schoolchildren (Böhning et al., 1999).*

Index	0	1	2	3	4	5	6	7	8
Frequency	172	73	96	80	95	83	85	65	48

Table 10.5 records the dental status of children according to an index that takes values 0–8. Develop an r^*-type approximation for the zero-inflation parameter ω and find the P-value for testing the non-regular hypothesis $\omega = 0$ (Castillo and López-Ratera, 2006).

Deng and Paul (2005) consider score tests for the more difficult problem of testing $\omega = 0$ in the presence of overdispersion. Higher order approximation for the overdispersed Poisson model is discussed in Davison *et al.* (2006).

Chapter 5

30. Investigate the effect of changing the degrees of freedom ν on the lengths of the confidence intervals found in analysis of the nuclear power data.
31. Sen and Srivastava (1990, p. 32) give data on the selling prices of 26 houses, available as data frame houses in the marg package. The variables considered as potentially related to the selling price (price) are number of bedrooms (bdroom), the floor space in square feet (floor), the total number of rooms (rooms) and the front footage of the lot in feet (front).

   (a) Fit a linear model relating price to the other variables using both lm and rsm, in the latter case allowing the errors to follow a Student t distribution.
   (b) Construct first order and third order confidence intervals for each regression coefficient in turn, for normal errors and for Student t_ν errors for $\nu = 3, 5, 7$. Investigate the stability of the conclusions to the assumption on the errors.
   (c) Plot the profile and modified profile log likelihood functions for each regression coefficient.

(d) Plot the modified profile log likelihood for $\log \sigma$, assuming Student t_ν errors with a range of values for ν. Compare this to the log likelihood for a $\sigma^2 \chi^2_{n-p}$ density, which is the exact marginal log likelihood under the normality assumption, and to the log likelihood for a $\sigma^2 \chi^2_n$ density, as would be obtained from profile log likelihood for normal errors.

32. Analyse Darwin's maize plant data, in `darwin` of package `marg`, using a t_4 distribution. Try fitting the model with pot effects and pairs effects. At what point does the fitting break down?

33. Under a linear model the responses Y_j are independent normal variables with means μ_j and variances σ^2; let E_g denote expectation with respect to this true but unknown model g. We aim to choose a candidate model $f(y; \theta)$ to minimize the difference of log likelihoods when predicting a new sample like the old one,

$$\Delta = \mathrm{E}_g \left(\mathrm{E}_g^+ \left[2 \sum_{j=1}^n \log \left\{ \frac{g(Y_j^+)}{f(Y_j^+; \widehat{\theta})} \right\} \right] \right), \tag{10.4}$$

where $Y_1^+, \dots, Y_n^+$ is another sample independent of $Y_1, \dots, Y_n$ but with the same distribution, E_g^+ denotes expectation over $Y_1^+, \dots, Y_n^+$, and $\widehat{\theta}$ is the maximum likelihood estimator of θ based on $Y_1, \dots, Y_n$.

If the candidate model is normal, then θ comprises the mean responses $\mu_1, \dots, \mu_n$ and σ^2, with maximum likelihood estimators $\widehat{\mu}_1, \dots, \widehat{\mu}_n$ and $\widehat{\sigma}^2$. Show that the sum in (10.4) equals

$$\frac{1}{2} \sum_{j=1}^n \left\{ \log \widehat{\sigma}^2 + \frac{(Y_j^+ - \widehat{\mu}_j)^2}{\widehat{\sigma}^2} - \log \sigma^2 - \frac{(Y_j^+ - \mu_j)^2}{\sigma^2} \right\},$$

and deduce that the inner expectation is

$$\sum_{j=1}^n \left\{ \log \widehat{\sigma}^2 + \frac{\sigma^2}{\widehat{\sigma}^2} + \frac{(\mu_j - \widehat{\mu}_j)^2}{\widehat{\sigma}^2} - \log \sigma^2 - 1 \right\}.$$

A candidate linear model with full-rank $n \times p$ design matrix X is correct if the true model is nested within it. Show that in this case, $\sum (\mu_j - \widehat{\mu}_j)^2 \sim \sigma^2 \chi^2_p$, independent of $n\widehat{\sigma}^2 \sim \sigma^2 \chi^2_{n-p}$. Hence show that if $n - p > 2$ then

$$\Delta = n\mathrm{E}_g \left(\log \widehat{\sigma}^2 \right) + \frac{n^2}{n-p-2} + \frac{np}{n-p-2} - n \log \sigma^2 - n.$$

Show that apart from a constant this is estimated unbiasedly by the *corrected information criterion*

$$\mathrm{AIC}_c = n \log \widehat{\sigma}^2 + n \frac{1 + p/n}{1 - (p+2)/n};$$

the 'best' candidate model is taken to be that which minimizes AIC_c.

Establish that $\text{AIC}_c \doteq n \log \widehat{\sigma}^2 + n + 2(p+1) + O(p^2/n)$, and hence for large n and fixed p this selects the same model as does $\text{AIC} = n \log \widehat{\sigma}^2 + 2p$, while AIC_c penalizes model dimension more severely when p is comparable with n.

(Hurvich and Tsai, 1989, 1991)

34. In the regression-scale model (5.1), an expression for the exact conditional density of $(\widehat{\beta}, \widehat{\sigma})$, given a (cf. (5.2)) can be obtained by a transformation of variables from $y = (y_1, \ldots, y_n)$ to $(a_1, \ldots, a_n, \widehat{\beta}, \widehat{\sigma})$, where there are $p+1$ constraints on the vector a from the maximum likelihood equations. The resulting expression is

$$f_1(\widehat{\beta}, \widehat{\sigma} \mid a; \beta, \sigma) = \frac{\widehat{\sigma}^{n-p-1}}{\sigma^n} \prod_{i=1}^{n} f \left\{ \frac{a_i \widehat{\sigma} + x_i^{\mathrm{T}}(\widehat{\beta} - \beta)}{\sigma} \right\} k^{-1}(a, n), \qquad (10.5)$$

where $f(\cdot)$ is the known error density, and $k(a, n)$ is the normalizing constant for the conditional density. The factor $\widehat{\sigma}^{n-p-1}$ comes from the Jacobian of the transformation from y to $(\widehat{\beta}, \widehat{\sigma}, a)$. Show that in the normal case $f(\varepsilon) = (2\pi)^{-1/2} \exp(-\varepsilon^2/2)$, for $\varepsilon \in \mathbb{R}$:

(a) the maximum likelihood estimates are

$$\widehat{\beta} = (X^{\mathrm{T}}X)^{-1}X^{\mathrm{T}}y, \qquad \widehat{\sigma}^2 = (y - X\widehat{\beta})^{\mathrm{T}}(y - X\widehat{\beta})/n;$$

(b) the constraints on the ancillary vector a are

$$\sum a_i x_i^{\mathrm{T}} = 0, \qquad \sum a_i = n; \text{ and}$$

(c) expression (10.5) simplifies to the product of a $N(\beta, \sigma^2(X^{\mathrm{T}}X)^{-1})$ density for $\widehat{\beta}$ and a $\sigma^2 \chi_{n-p}^2$ density for $n\widehat{\sigma}^2$.

Conditional inference in linear regression is discussed in Fraser (1979, Chapter 6) and Lawless (2003, Appendix E). The basis for the tail area approximations in the `marg` package is set out in DiCiccio *et al.* (1990), building on the density expressions in Barndorff-Nielsen (1980).

35. In the factorization of (5.2) it is not necessary that $\widehat{\beta}$ and $\widehat{\sigma}$ be the maximum likelihood estimators. Show that (5.2) holds for any *equivariant* estimators $(\widetilde{\beta}, \widetilde{\sigma})$, i.e. estimators satisfying

$$\widetilde{\beta}\{(y - X\beta)/\sigma\} = \{\widetilde{\beta}(y) - \beta\}/\sigma, \qquad \widetilde{\sigma}\{(y - X\beta)/\sigma\} = \widetilde{\sigma}(y)/\sigma;$$

the least squares estimators are often a convenient choice, even when the underlying density is non-normal.

36. The heteroscedastic nonlinear regression model is

$$y_{ij} = \mu(x_i; \beta) + w(x_i; \beta, \rho)\varepsilon_{ij}, \quad i = 1, \ldots m, \quad j = 1, \ldots, n_i, \tag{10.6}$$

where m is the number of design points, n_i the number of replicates at design point x_i, the errors ε_i are independent standard normal random variables, and β and ρ may be scalar or vector. This model can be embedded in a curved exponential family of order $(2m, d)$, where d is the dimension of the parameter $\theta = (\beta, \rho)$ and m is the number of design points. Show that the canonical parameter of the embedding full family is

$$\phi(\theta) = \left(\frac{\mu(x_1; \beta)}{w(x_1; \beta, \rho)^2}, \ldots, \frac{\mu(x_m; \beta)}{w(x_m; \beta, \rho)^2}, -\frac{1}{2w(x_1; \beta, \rho)^2}, \ldots, -\frac{1}{2w(x_m; \beta, \rho)^2} \right)^{\mathrm{T}},$$

and the corresponding sufficient statistic is

$$t(y) = \left(\sum_{j=1}^{n_1} y_{1j}, \ldots, \sum_{j=1}^{n_m} y_{mj}, \sum_{j=1}^{n_1} y_{1j}^2, \ldots, \sum_{j=1}^{n_m} y_{mj}^2 \right)^{\mathrm{T}},$$

with covariance matrix

$$\Sigma(\theta) = \begin{pmatrix} \mathrm{diag}(n_i w_i^2) & \mathrm{diag}(2n_i w_i^2 \mu_i) \\ \mathrm{diag}(2n_i w_i^2 \mu_i) & \mathrm{diag}(2n_i w_i^4 + 4w_i^2 \mu_i^2) \end{pmatrix},$$

where $\mu_i = \mu(x_i; \beta)$ and $w_i = w(x_i; \beta, \rho)$. Furthermore, derive the cumulant function

$$K(\theta) = -\sum_{i=1}^{m} \left\{ \frac{n_i \phi_{1i}^2}{4\phi_{2i}} + \frac{n_i}{2} \log(-2\phi_{2i}) \right\}$$

of the embedding model, where ϕ_1 and ϕ_2 represent respectively the first and second m elements of the vector ϕ. This is the basis for Skovgaard's approximation to q_1, given at (8.40), which is implemented in the `nlreg` package. If the errors are not assumed to be normal, then Skovgaard's approach is not directly available, although the approach based on q_2 given at (8.33) is. The ancillary directions V are constructed using the standardized residuals.

A technical point arises in the normal setting: the embedding curved exponential family has dimension $2m$ and the asymptotic theory applies to embedding exponential families of fixed dimension. Bellio (2000, Section 2.1.1) indicates that this poses no problem if we require only second order accuracy. If m is fixed but $n_i \to \infty$ the problem does not arise.

37. The data in Table 10.6 are taken from Cox and Snell (1981, Example D), where they are used to illustrate a non-standard application of simple linear regression. Temperature t_{ME} is measured in each of 20 sections of a chemical reactor, and this

Table 10.6 *Measured and theoretical temperatures (°C) in 20 sections of a reactor (Cox and Snell, 1981, Example D).*

Measured	Theoretical	Measured	Theoretical
431	432	472	498
450	470	465	451
431	442	421	409
453	439	452	462
481	502	451	491
449	445	430	416
441	455	458	481
476	464	446	421
460	458	466	470
483	511	476	477

is compared to both the theoretical temperature, t_{TH}, and the true temperature t_{TRUE} via the model

$$t_{TRUE} = \alpha + \beta(t_{TH} - \overline{t_{TH}}) + \epsilon,$$

$$t_{ME} = t_{TRUE} + \varepsilon,$$

where $\epsilon \sim N(0, \sigma^2)$ is a random error term, ε is a measurement error known to follow a normal distribution with mean 0 and variance 9, $\overline{t_{TH}}$ is the mean theoretical temperature in the measured channels, and t_{TH} can be calculated for an effectively infinite population of sections, and is thus known to have mean 452°C and standard deviation 22°C. The parameter of interest is the probability of a value of t_{TRUE} greater than 490°C, i.e.

$$\psi = \Phi[\{\alpha + \beta(452 - \overline{t_{TH}}) - 490\}/\sqrt{(484\beta^2 + \sigma^2 - 9)}],$$

since t_{TRUE} follows a normal distribution with mean $\alpha + \beta(452 - \overline{t_{TH}})$ and variance $484\beta^2 + \sigma^2 - 9$. Cox and Snell (1981) compute an approximate confidence limit for $\Phi^{-1}(\psi)$ using a delta method approximation to the variance of the estimated numerator and denominator, leading to a point estimate of ψ of 0.0416, with a 97.5% upper confidence limit of 0.0226. Compare this with the results obtained using the normal approximation to the distribution of r^*.

38. Annis (2005, Table 1) considers a variant of Box's (1992) paper helicopter problem. The goal is to find the setting of the factors wing length (L) and width (W) that maximizes flight time T when a paper helicopter is dropped from a height of 15 inches. The length (B) and height (H) of the helicopter base are fixed to respectively 3 and 2 inches. The admissible values for wing length and width are restricted to

$2 \leq L \leq 6$ and $0.5 \leq W \leq 3$ inches. Annis (2005) suggests using a nonlinear model of the form

$$E(Y) = \beta_0 + \beta_1 \log \left(\frac{\beta_2^2}{LW} + LW \right), \tag{10.7}$$

where the response variable $Y = \log(T) + \log(S)/2$ depends on the surface area of the helicopter $S = BH + (2L + 1)W$.

(a) Fit (10.7) using the data `helicopter` and the `nlreg` routine of package `nlreg` and verify that the estimates $\widehat{\beta}_0 = 5.54$, $\widehat{\beta}_1 = -0.69$ and $\widehat{\beta}_2 = 20.47$ are all statistically significant at the 5% level.
(b) Using the `profile` method produce the first order and higher order profile plots. Is there any need for higher order inference?
(c) Show that for the above estimates, flight time is maximized if $(L, W) = (6.0, 1.77)$ inches.

39. In assays which measure a biological endpoint, the variability of the response often changes with its level. Carroll and Ruppert (1988, Section 2.8) discuss several graphical techniques to detect such variance heterogeneity. Departure from homoscedasticity may be accommodated by using the nonlinear model (10.6) which includes the variance function $w(x_i; \beta, \rho)^2$. A second possibility is to seek a transformation of the data, $h(y; \lambda)$, indexed by the parameter λ, such that the transformed response has constant variance (Carroll and Ruppert, 1988, Chapter 4). To maintain the meaning of the parameters in the nonlinear mean function $\mu(x_i; \beta)$, model (10.6) is replaced by the *transform-both-sides* model

$$h(y_{ij}; \lambda) = h\{\mu(x_i; \beta); \lambda\} + \sigma \varepsilon_{ij}, \quad i = 1, \ldots, m, \quad j = 1, \ldots, n_i. \tag{10.8}$$

The most commonly used family is the modified power transformation family (Box and Cox, 1964)

$$h(y; \lambda) = \begin{cases} \dfrac{y^\lambda - 1}{\lambda}, & \lambda \neq 0, \\ \log y, & \lambda = 0, \end{cases} \quad \lambda \in \mathbb{R}.$$

The parameter λ may be estimated from the data or may be known from previous analyses.

(a) The special case $\lambda = 0$ corresponds to log transformation of both sides of (10.6), and has an appealing interpretation: the data on the original scale,

$$y_{ij} = \mu(x_i; \beta) \exp(\sigma \varepsilon_{ij}),$$

have a multiplicative error structure with errors that follow a log-normal distribution. Show that in this case $\mu(x_i; \beta)$ represents the median response. What is the mean response?

(b) A frequently used variance function is the *power-of-the-mean* variance function $w(x_i; \beta, \rho)^2 = \sigma^2 \mu(x_i; \beta)^\rho$, where ρ is a real parameter. Use Taylor series expansion to show that model (10.6) is equivalent to model (10.8) to the first order, where $h'(\cdot)$ is proportional to $w^{-1}(\cdot)$.   (Carroll and Ruppert, 1988, p. 121)

(c) Growth data often exhibit constant coefficient of variation rather than constant variance. Show that in this case $w(x_i; \beta, \rho) = \sigma \mu(x_i; \beta)$, and that the variance can be stabilized by using a transform-both-sides model with $h(\cdot) = \log(\cdot)$.

40. The four-parameter logistic model with threshold x_0 is

$$\mu(x; \beta) = \beta_1 + \frac{\beta_2 - \beta_1}{1 + \{(x - x_0)_+ / \beta_4\}^{2\beta_3}}, \quad x, x_0 \geq 0, \ \beta_1, \ldots, \beta_4 \geq 0,$$

where $(x - x_0)_+ = \max(x - x_0, 0)$.

(a) Show that the interpretation of the parameters β_1, β_2 and β_3 is the same as for the standard logistic model (5.5), while the EC_{50} equals $\beta_4 + x_0$.

(b) The code below implements the iterative procedure which calculates and plots the profile log likelihood function for the threshold value x_0, for the `daphnia` data analyzed in Section 5.3.

```
log.time <- log(daphnia$time)
x0.seq <- seq(0, 0.2, 0.01)
logLik.x0 <- matrix(0, nrow = length(x0.seq), ncol = 2)
for( idx in seq(along = x0.seq) )

    x0 <- x0.seq[idx]
    conc.x0 <- ifelse(daphnia$conc - x0 > 0, daphnia$conc - x0, 0)
    fitted.mod <-
        nlreg( log.time ~ log(b1+(b2-b1)/(1+(conc.x0/b4)^(2*b3))),
               start = list(b1 = 0.05, b2 = 50, b3 = 3, b4 = 0.1),
               hoa = TRUE )
    logLik.x0[idx,] <- c( x0, logLik(fitted.mod) )

    plot( logLik.x0, type = 'l', ylab = 'profile log likelihood'
          xlab = expression(paste('threshold ', x[0])))
```

Using the values in `logLik.x0`, calculate the maximum likelihood estimate, its standard error and a 95% confidence interval based on the Wald pivot for the threshold parameter x_0.

Chapter 6

41. Find the elements of the local exponential family approximation used in (2.11) for a model for binomial response data with parametrized link function (6.1), where $\log \xi = x^T \beta$.

42. The complementary log–log link for binary data arises in serial dilution assays, which are used to estimate the density of organisms per unit of volume of solution when it is possible only to determine the presence or absence of the organism in a sample. Suppose that at the ith dilution n_i samples of volume $v_i > 0$ are tested, of which y_i

are found to be positive for presence of the organism. Under the assumption that the organisms are randomly distributed, the probability that the jth sample for dilution i is positive equals $p_i = 1 - \exp(-\lambda v_i)$, where $\lambda > 0$ is the density of organisms per unit volume.

(a) Show that on the basis of k dilutions with n_i samples at dilution v_i, the log likelihood function for λ is that of a binomial distribution with complementary log–log link, with $\log \lambda$ the intercept and fixed offset $\log v_i$, for $i = 1, \ldots, k$.

(b) Ridout (1994, Table 1) compares confidence intervals based on likelihood methods to some exact methods, for $k = 3$, $n_i = 3$, $v_i = 0.1^i$, and several outcomes (y_1, y_2, y_3). Using the higher order approximation described in Section 6.2, compare the first and higher order methods.

(c) By considering all possible outcomes the actual coverage probability of the intervals from (b) can be compared (Ridout, 1994, Figure 1). Compare the coverage using r^* with that from first order likelihood inference.

The outcomes $(0, 0, 0)$ and $(1, 1, 1)$ lead to $\widehat{\lambda}$ either 0 or ∞, so the asymptotic theory does not apply. An ad hoc correction is suggested in Ridout (1994). The Poisson model is generalized in Stallard *et al.* (2006) in the context of testing infectivity of tissue samples, motivated by a problem of testing for transmissible spongiform encephalopathies, such as BSE.

43. Cook and Weisberg (1982, Section 2.2.2) suggest studying whether the jth observation in model (5.1) is an outlier by using mean shift outlier model

$$y_i = x_i^{\mathrm{T}}\beta + \Delta\delta_j + \sigma\varepsilon_i, \quad i = 1, \ldots, n,$$

where δ_j is the indicator variable selecting the jth case. Nonzero values of Δ imply that observation j is an outlier. Bellio (2000) extends this framework to the nonlinear models (5.3) and (5.4). Deletion residuals and r^*-type deletion residuals are defined as respectively the likelihood root and the modified likelihood root for testing $\Delta = 0$. The latter is computed in the `nlreg` package using Skovgaard's approximation. Furthermore, according to Cook and Weisberg (1982, Section 5.2) we may define Cook's distance for nonlinear models as

$$C_j = \frac{2}{d}\left\{\ell(\widehat{\theta}) - \ell(\widehat{\theta}_{-j})\right\},$$

where $\ell(\theta)$ is the log likelihood for the nonlinear model parametrized by the d-dimensional parameter $\theta = (\beta, \rho)$, $\widehat{\theta}$ is the maximum likelihood estimate, and $\widehat{\theta}_{-j}$ is the maximum likelihood estimate after deleting the jth case. The influence measure C_j is a global influence measure and does not distinguish between regression coefficients and variance parameters, for which purpose Bellio (2000) proposes the use of

$$C_j(\beta \mid \rho) = \frac{2}{p}\left\{\ell_{\mathrm{p}}(\widehat{\beta}) - \ell_{\mathrm{p}}(\widehat{\beta}_{-j})\right\}, \quad C_j(\rho \mid \beta) = \frac{2}{d-p}\left\{\ell_{\mathrm{p}}(\widehat{\rho}) - \ell_{\mathrm{p}}(\widehat{\rho}_{-j})\right\},$$

where p is the dimension of β and ℓ_p is the profile log likelihood.

Use the `nlreg.diag.plots` routine of the `nlreg` package to compute the regression diagnostics for the models fitted to the chlorsulfuron herbicide data C1–C4 in Section 6.4. Are there any outlying and/or influential observations?

44. In a series of herbicide experiments Rudemo *et al.* (1989) observed that the variance of the residuals from standard model fits such as (5.3), (5.4) and (10.8) was largest where the derivative $d\mu(x; \beta)/dx$ of the mean response with respect to the covariate x was largest. This led them to postulate a Berkson-type measurement error model (Berkson, 1950)

$$x^{\text{true}} = x + \sigma_x \nu,$$

where x^{true} is the true but unobserved value and x is the observed proxy measure. The error term $\sigma_x \nu$ follows a centred normal distribution and its variance may depend on the covariate x. One interpretation of this model is that x represents the nominal herbicide dose, while x^{true} is the actual dose absorbed by the plants.

Assume for the dose-response model the transform-both-sides model

$$h(y; \lambda) = h\{\mu(x^{\text{true}}; \beta); \lambda\} + \sigma \varepsilon,$$

where ε is standard normal. Assume ε and ν are independent. Show by first order Taylor series expansion around x that

$$h(y; \lambda) \simeq h\{\mu(x; \beta); \lambda\} + g(x; \beta, \lambda)\xi,$$

where ξ is a standard normal error and

$$g^2(x; \beta, \lambda) = \sigma^2 \left[1 + \frac{\sigma_x^2}{\sigma^2} \left\{ \frac{dh(y; \lambda)}{dy} \frac{d\mu(x; \beta)}{dx} \right\}^2 \right].$$

If $h(\cdot)$ is the modified power transformation (Box and Cox, 1964)

$$h(y; \lambda) = \begin{cases} \dfrac{y^\lambda - 1}{\lambda}, & \lambda \neq 0, \\ \log y, & \lambda = 0, \end{cases} \quad \lambda \in \mathbb{R},$$

and $\sigma_x = \sigma_1 x^{\gamma/2}$, $\kappa = \sigma_1^2/\sigma^2$, show that

$$g^2(x; \beta, \lambda, \gamma, \kappa) = \sigma^2 \left[1 + \kappa x^\gamma \left\{ \mu(x; \beta)^{\lambda-1} \frac{d\mu(x; \beta)}{dx} \right\}^2 \right].$$

Chapter 7

45. Suppose that

$$y_{rt} = \mu_t + \alpha + \eta_r + \varepsilon_{rt}, \quad r = 1, \ldots, n, \ t = 1, \ldots, T,$$

where α, η_r and ε_{rt} are mutually independent normal random variables with zero means and variances σ_α^2, σ_η^2 and σ^2, and the μ_t are regarded as fixed unknown parameters. Show that the quantities

$$y_{rt} - \bar{y}_{r.} - \bar{y}_{.t} + \bar{y}_{..}, \quad \bar{y}_{r.} - \bar{y}_{..}, \quad \bar{y}_{.t} - \bar{y}_{..}, \quad \bar{y}_{..}, \quad r = 1, \ldots, n, \ t = 1, \ldots, T,$$

are independent normal random variables and that we can write

$$SS = \sum_{r,t}(y_{rt} - \bar{y}_{r.} - \bar{y}_{.t} + \bar{y}_{..})^2 = \sum_{r,t}(\varepsilon_{rt} - \bar{\varepsilon}_{r.} - \bar{\varepsilon}_{.t} + \bar{\varepsilon}_{..})^2,$$

$$SS^R = \sum_{r,t}(\bar{y}_{r.} - \bar{y}_{..})^2 = T\sum_{r}(\eta_r + \bar{\varepsilon}_{r.} - \bar{\eta}_{.} - \bar{\varepsilon}_{..})^2,$$

$$SS^T = \sum_{r,t}(\bar{y}_{.t} - \bar{y}_{..})^2 = n\sum_{t}(\mu_t - \bar{\mu}_{.} + \bar{\varepsilon}_{.t} - \bar{\varepsilon}_{..})^2.$$

Deduce that these are independent with distributions

$$SS \sim \sigma^2 \chi^2_{(T-1)(n-1)}, \quad SS^R \sim (\sigma^2 + 4\sigma_\eta^2)\chi^2_{n-1}, \quad SS^T \sim \sigma^2 \chi^2_{T-1}(\delta),$$

where the last is non-central chi-square with non-centrality parameter $\delta = \sum(\mu_t - \bar{\mu}_{.})^2$, and that independently of these,

$$\bar{y}_{..} \sim N\{\bar{\mu}_{.}, \sigma_\alpha^2 + \sigma_\eta^2/n + \sigma^2/(nT)\}.$$

46. In Problem 34 we showed that in the normal theory linear model the residual sum of squares $(y - X\hat{\beta})^T(y - X\hat{\beta})$ follows a $\sigma^2 \chi^2_{n-p}$ distribution, independently of the least squares estimator $\hat{\beta}$. To use this marginal density for likelihood-based inference on σ^2, we have

$$\ell_m(\sigma^2) = -\frac{n-p}{2}\log\sigma^2 - \frac{1}{2\sigma^2}\hat{\varepsilon}^T\hat{\varepsilon},$$

where $\hat{\varepsilon} = y - X\hat{\beta}$. The maximum marginal log likelihood estimator of σ^2 is $\hat{\sigma}_m^2 = \hat{\varepsilon}^T\hat{\varepsilon}/(n-p)$, and the χ^2 approximation to its distribution is exact.

The construction of a log likelihood based on residuals has been extensively generalized under the name of REML, for restricted (or residual) maximum likelihood. It is widely used in analysis of the linear mixed model

$$y = X\beta + Zb + \varepsilon,$$

where X and Z are known $n \times p$ and $n \times k$ matrices, β is a $p \times 1$ vector of fixed effects and b is a $k \times 1$ vector of random effects. We assume that $\varepsilon \sim N_n(0, \sigma^2 \Lambda)$, and $b \sim N_k(0, \sigma^2 \Psi)$, and that ε and b are independent, and we denote by ρ the q-dimensional vector of distinct parameters on which the matrix $\sigma^{-2} V = Z \Psi Z^{\mathrm{T}} + \Lambda$ depends.

(a) Show the constrained maximum likelihood estimate of β for fixed ρ and σ^2 is the usual weighted least squares estimator

$$\widehat{\beta}_\rho = (X^{\mathrm{T}} V X)^{-1} X^{\mathrm{T}} V y.$$

(b) The REML log likelihood is a function of the residuals $\widehat{\varepsilon} = y - X\widehat{\beta}_\rho$, and takes the form

$$\ell_{\mathrm{R}}(\rho, \sigma^2) = -\frac{1}{2}\log|V| - \frac{1}{2}\log|X^{\mathrm{T}} V X| - \frac{1}{2}\widehat{\varepsilon}^{\mathrm{T}} V^{-1} \widehat{\varepsilon}.$$

Show that this is equal to the marginal log likelihood of $U = B^{\mathrm{T}} y$, where B is any $n \times (n - p)$ matrix satisfying $BB^{\mathrm{T}} = I - X(X^{\mathrm{T}} X)^{-1} X^{\mathrm{T}}$ and $B^{\mathrm{T}} B = I$ (Davison, 2003, Section 12.2).

In as-yet unpublished work R. Bellio and A. R. Brazzale discuss an alternative to REML based on adjusted profile likelihood for generalized linear mixed models.

47. The expectations needed to compute S and Q for Skovgaard's approximation to q_1 in the mixed linear model

$$y_i = X_i \beta + Z_i b_i + \varepsilon_i, \quad i = 1, \ldots, m,$$

where y_i is $n_i \times 1$, X_i and Z_i are respectively $n_i \times p$ and $n_i \times q$ matrices of constants, $b_i \sim N(0, \psi)$, and $\varepsilon_i \sim N(0, \Lambda_i)$, are given explicitly in Lyons and Peters (2000), for independent within-group errors ε_i; that is, for $\Lambda_i = \sigma^2 I_{n_i}$. Guolo and Brazzale (2005) extend this to the more general linear mixed model with arbitrary Λ_i, hence accounting for heteroscedastic and correlated within-group errors. They compare inference based on normal approximation of the full and REML likelihood roots to Skovgaard's approximation based on the full log likelihood. Simulations indicate that inference based on r^* using Skovgaard's approximation to q_1 are quite accurate for inference about both fixed effects and parameters in the covariance matrix. Use of the REML likelihood root is essentially equivalent to making a nuisance parameter correction, but not an information correction. An extension of this work to the nonlinear mixed effects model is discussed in Guolo *et al.* (2006). Skovgaard's approximation is applied to the REML log likelihood in the unpublished PhD dissertation of Mette Krog-Josiassen.

48. The Wishart distribution is the multivariate generalization of the χ^2 distribution. A symmetric positive definite matrix S follows a $\text{Wishart}_p(n, \Omega)$ distribution if its density is given by

$$f_S(s; \Omega) = \frac{|s|^{(n-p-1)/2} \exp\{-\text{tr}(\Omega^{-1}s)/2\}}{2^{np/2}|\Omega|^{n/2}\Gamma_p(n/2)}, \quad n \geq p,$$

where $\Gamma_p(n/2) = \pi^{p(p-1)/4} \prod_{j=1}^{p} \Gamma\{(n+1-j)/2\}$, and Ω is a symmetric positive definite matrix. Typically S is a matrix of sums of squares and cross-products for a normal random vector. Suppose that S_e and S_h are two such matrices, where

$$S_e \sim \text{Wishart}_p(n, \Sigma_e), \quad S_h \sim \text{Wishart}_p(q, \Sigma_h),$$

and S_e and S_h are independent. One test of $\Sigma_e = \Sigma_h$ is based on the *Bartlett–Nanda–Pillai trace statistic* $V = \text{tr}\{S_e/(S_e + S_h)\}$, small values of which represent evidence against equality of the covariance matrices. In multivariate analysis of variance this arises as a test of equality of mean vectors among groups. Butler *et al.* (1992b) show how this problem can be embedded in an exponential family model.

Write $\Sigma_h^{-1} = \Delta$, say, where $\Delta = (\delta_{ij})$ is a $p \times p$ symmetric, positive definite matrix, and define the scalar parameter β by $\Sigma_e^{-1} = \Delta - 2\beta I_p$; the values (β, Δ) are constrained to ensure that Σ_e^{-1} is positive definite. Show that the log likelihood function for β, δ is that of a linear exponential family:

$$\ell(\beta, \delta) = \beta t + \delta^{\text{T}} u - K(\beta, \delta),$$

where

$$\delta = (-\tfrac{1}{2}\delta_{11}, \ldots, \tfrac{1}{2}\delta_{pp}, -\delta_{12}, \ldots, -\delta_{p-1,p}),$$
$$u = \{(S_e + S_h)_{qq}, \ldots, (S_e + S_h)_{pp}, (S_e + S_h)_{12}, \ldots, (S_e + S_h)_{p-1,p}\},$$
$$K(\beta, \delta) = \tfrac{1}{2}n \log|\Delta - 2\beta I_p| - \tfrac{1}{2}q \log|\Delta|,$$

and $T = \text{tr}(S_e)$. Hence show that

$$\Pr(V \leq t; \beta = 0) = \Pr(T \leq t \mid U = u; \beta = 0);$$

this is the basis for approximating the null distribution of V. Butler *et al.* (1992b) provide extensive discussion of the approximation, using both the r^* approximation based on the profile log likelihood and that based on the conditional log likelihood; cf. Section 8.6.1.

Chapter 8

49. Verify that $\ell_p(\psi)$ is maximized at $\psi = \widehat{\psi}$, the ψ-component of the full maximum likelihood estimator, and that $j_p(\widehat{\psi})^{-1}$ is an estimate of the asymptotic covariance of $\widehat{\psi}$. (Section 8.2)

50. (a) Assume that y can be expressed explicitly as a one-to-one function of $(\widehat{\theta}, a)$, where $\widehat{\theta}$ is the maximum likelihood estimator. Differentiate the score equation to verify that $\ell_{\theta;\widehat{\theta}}(\widehat{\theta}; \widehat{\theta}, a) = j(\widehat{\theta})$.

(b) Assume that y is a one-to-one function of $\widehat{\theta}$, and there is no need for a supplementary statistic a. This is the case in a full exponential family model, or in any scalar parameter model with a single observation. Show that $\ell_{;\widehat{\theta}}(\theta; \widehat{\theta}) = \ell_{;y}(\theta; \widehat{\theta})\,(\partial y/\partial \theta)|_{\widehat{\theta}}$.

(c) Show that the same identity as in part (b) holds for a sample of size n from a scalar parameter location model, where the differentiation with respect to $\widehat{\theta}$ is for fixed $(a_1, \ldots, a_n)$, where $a_i = y_i - \widehat{\theta}$.

(d) Assume that y is a sample of size n from a two-parameter exponential family model $f(y_i; \theta) = \exp\{\psi u(y_i) + \lambda v(y_i) - c(\theta) + h(y_i)\}$. Show that $\ell_{;y}(\theta; y) = \sum_{i=1}^{n} \partial \log\{f(\theta; y_i)\}/\partial y_i$, evaluated at a fixed point y^0, is affinely equivalent to the canonical parameter (ψ, λ).

(Section 8.2)

51. For a scalar parameter of interest ψ, use (8.9) and an expansion of the left-hand side of $\ell_\lambda(\psi, \widehat{\lambda}_\psi) = 0$ to obtain the normal approximation to $t(\psi)$ given below (8.11). (Section 8.3)

52. For a sample of size n from the Cauchy density

$$f(y; \theta) = \frac{1}{\pi\{1 + (y - \theta)^2\}}, \quad y, \theta \in \mathbb{R},$$

compute the observed and expected Fisher information. Carry out some simulations to show that $\mathrm{var}\{\widehat{\theta} \mid j(\widehat{\theta})\} \doteq j(\widehat{\theta})^{-1}$. This is one of the examples discussed in Efron and Hinkley (1978), who show that $\mathrm{var}\{\widehat{\theta} \mid j(\widehat{\theta})\}$ is a good approximation to $\mathrm{var}(\widehat{\theta} \mid a)$, where a is the usual ancillary statistic for the location model. (Section 8.4)

53. In the p^* density approximation (8.12), show that Taylor series approximation of the log likelihood function leads to a normal approximation for $\widehat{\theta}$ with mean θ and variance $j(\widehat{\theta})$. This and the result of Problem 52 motivate the recommendation that $j(\widehat{\theta})$ is the best choice for standardizing the maximum likelihood estimator, among asymptotically equivalent versions $i(\theta)$, $i(\widehat{\theta})$ and $j(\theta)$. (Section 8.4)

54. The general form for the canonical parameter of the tangent exponential model for nonlinear regression is given at (8.18). In this problem we investigate several special cases.

(a) *Normal errors and non-constant variance*: with $f_0(\cdot) = \phi(\cdot)$ we have

$$\varphi(\theta)^{\mathrm{T}} = -1_n^{\mathrm{T}} \mathcal{E} W^{-1} \left(\widehat{X}_\beta + \widehat{\mathcal{E}} \widehat{W}_\beta \quad \widehat{\mathcal{E}} \widehat{W}_\rho \right),$$

where 1_n is a n-dimensional vector of ones.

(b) If furthermore $w(x_i; \beta, \rho) = \sigma$ is constant, the canonical parameter for the classical nonlinear regression model is

$$\varphi(\theta)^{\mathrm{T}} = -\frac{1}{\sigma} 1_n^{\mathrm{T}} \mathcal{E} \left(\widehat{X}_\beta \quad \widehat{\mathcal{E}} \right).$$

(c) *Linear regression with arbitrary errors*: if in (8.17) $\mu(x_i; \beta) = x_i^{\mathrm{T}} \beta$ and the variance is constant, we have

$$\varphi(\theta)^{\mathrm{T}} = -\frac{1}{\sigma} g_1^{\mathrm{T}} \left(X \quad \widehat{\mathcal{E}} \right),$$

which simplifies to

$$\varphi(\theta)^{\mathrm{T}} = -\frac{1}{\sigma} 1_n^{\mathrm{T}} \mathcal{E} \left(X \quad \widehat{\mathcal{E}} \right)$$

in the normal case and yields expression (8.16) for a location family model.

(Section 8.4)

55. The canonical parameter of the tangent exponential model is derived by finding some vectors V tangent to an approximate ancillary statistic. This argument fails for count data, as the sample space or conditional sample space is a lattice. Davison *et al.* (2006) argue that it is appropriate to use $\mu_i = E(y_i)$ in place of y_i in the definition of V. Show that this leads to $V = (1, -1, -1, 1)$ in the curved exponential family of Illustration 8.4.2. (Section 8.4)

56. Use the assumptions that r and q are related as indicated in expansion (8.23), and that the tail area approximation defines a proper cumulative distribution function, to verify that the normalizing constant c in the p^* approximation is $\sqrt{(2\pi)}\{1 + O(n^{-1})\}$. The normalizing constant for p^* is related to the Bartlett correction factor; see Barndorff-Nielsen and Cox (1994, Section 6.5). An explicit expression for regression-scale models is also given by DiCiccio *et al.* (1990). (Section 8.5)

57. With r and q given by (8.26) and (8.27), respectively, show that r has an expansion in terms of q of the form (8.23). Note that differentiation of the score equation $\ell_\theta(\widehat{\theta}; \widehat{\theta}, a) = 0$ leads to $j(\widehat{\theta}) = \ell_{\theta;\widehat{\theta}}(\widehat{\theta}; \widehat{\theta}, a)$ (Problem 50). A generalization of this result to higher order derivatives, and expansions of likelihood quantities in terms of mixed derivatives of ℓ, are discussed in Barndorff-Nielsen and Cox (1994, Chapter 5) and Severini (2000a, Section 6.2). Barndorff-Nielsen and Cox (1994, Section 6.7) show that $r^{-1} \log(q/r) \to 0$ and $q/r^3 \to 0$ as $\widehat{\theta} \to \pm\infty$, and that as $\widehat{\theta} \to \theta$, $r/u \to 1 + \{\ell_{\theta\theta\theta}(\widehat{\theta}; \widehat{\theta}, a) + 3\ell_{\theta\theta;\widehat{\theta}}(\widehat{\theta}; \widehat{\theta}, a)\} j(\widehat{\theta})^{-3/2}$, cf. (A.16). (Section 8.5)

58. Write the log likelihood function for a sample of size n from the linear exponential family (8.44) as a function of $(\widehat{\psi}, \widehat{\lambda})$, and use (8.32) to show that $q_1 = q_2$ in this model. (Section 8.5)

59. In a sample from the linear exponential family (8.44), show that the conditional density of u given v is independent of λ. Apply the saddlepoint approximation for sums given at (A.4) of the Appendix to the densities of (u, v) and of v to show that

$$f(u \mid v; \psi) \doteq \exp\{\ell_{\mathrm{p}}(\psi) - \ell_{\mathrm{p}}(\widehat{\psi})\} j_{\mathrm{p}}(\widehat{\psi})^{-1/2} \frac{|j_{\lambda\lambda}(\psi, \widehat{\lambda}_\psi)|^{1/2}}{|j_{\lambda\lambda}(\widehat{\psi}, \widehat{\lambda})|^{1/2}} \frac{c_1}{c_2}, \tag{10.9}$$

where $c_1/c_2 = 1/(2\pi)\{1 + O(n^{-1})\}$. The argument outlined in Section 8.5.2 can be applied to this expression: the factor $|j_{\lambda\lambda}(\psi, \widehat{\lambda}_\psi)|^{1/2}/|j_{\lambda\lambda}(\widehat{\psi}, \widehat{\lambda})|^{1/2}$ will be incorporated into the definition of q. Show that this leads to the usual tail area approximation, with r and q as defined at (8.45). (Section 8.6.1)

An alternative approach to approximating the conditional distribution incorporates the information adjustment into the log likelihood function. Write

$$\ell_{\mathrm{m}}(\psi) = \ell_{\mathrm{p}}(\psi) + \frac{1}{2} \log |j_{\lambda\lambda}(\widehat{\theta}_\psi)|;$$

then (10.9) becomes

$$f(u \mid v; \psi) \doteq c \exp\{\ell_{\mathrm{m}}(\psi) - \ell_{\mathrm{m}}(\widehat{\psi})\} j_{\mathrm{p}}(\widehat{\psi})^{-1/2}.$$

Now, using the results $\widehat{\psi}_{\mathrm{m}} = \widehat{\psi}\{1 + O(n^{-1})\}$ and $j_{\mathrm{m}}(\widehat{\psi}_{\mathrm{m}}) = j_{\mathrm{p}}(\widehat{\psi})\{1 + O(n^{-1})\}$ the same integration technique can be used to derive the tail area formulae (8.21) and (8.22) with r and q given below (8.50). This latter approximation is called the sequential saddlepoint approximation in Fraser *et al.* (1991) and comparisons between the two versions are discussed in Reid (1996), Butler *et al.* (1992b) and Pierce and Peters (1992).

60. In the linear regression model (8.48), show that if the parameter of interest is σ, the expression for q_2 simplifies to

$$\left(\frac{1}{\sigma} \tilde{g}_1^{\mathrm{T}} \widehat{\varepsilon} - \frac{n}{\widehat{\sigma}}\right) \left|\frac{1}{\sigma^2} X^{\mathrm{T}} \tilde{G}_2 X\right|^{1/2} |j(\widehat{\theta})|^{-1/2}.$$

(Section 8.6)

61. The simplification of $q_1 = q_2$ at (8.49) is given in the discussion of Pierce and Peters (1992). Verify this expression in the location-scale model $y_i = \mu + \sigma \varepsilon_i$, either directly or by simplifying the expression in the previous question using $X = (1, \ldots, 1)$.

(Section 8.6)

62. The components needed to construct q_2 in the nonlinear regression model given at (8.17) are outlined in Section 8.6.3. Verify the following special cases:

(a) *Constant variance*:

$$\varphi(\widehat{\beta}, \widehat{\rho}) = -\frac{1}{\widehat{\sigma}} \begin{pmatrix} 0_p \\ n \end{pmatrix}, \quad \varphi(\tilde{\beta}, \tilde{\rho}) = -\frac{1}{\widehat{\sigma}} \begin{pmatrix} \widehat{X}_\beta^{\mathrm{T}} \\ 1_n^{\mathrm{T}} \widehat{\varepsilon} \end{pmatrix} \tilde{g}_1$$

and

$$\varphi_{\theta^{\mathrm{T}}}(\tilde{\beta}, \tilde{p}) = \frac{1}{\tilde{\sigma}^2} \begin{pmatrix} \widehat{X}_\beta^{\mathrm{T}} \\ 1_n^{\mathrm{T}} \widehat{\mathcal{E}} \end{pmatrix} \begin{pmatrix} \tilde{G}_2 \tilde{X}_\beta & (\tilde{G}_2 \tilde{\mathcal{E}} + \tilde{G}_1) 1_n \end{pmatrix}.$$

(b) *Constant variance and linear mean function:*

$$\varphi(\widehat{\beta}, \widehat{p}) = -\frac{1}{\widehat{\sigma}} \begin{pmatrix} 0_p \\ n \end{pmatrix}, \quad \varphi(\tilde{\beta}, \tilde{p}) = -\frac{1}{\tilde{\sigma}} \begin{pmatrix} X^{\mathrm{T}} \\ 1_n^{\mathrm{T}} \widehat{\mathcal{E}} \end{pmatrix} \tilde{g}_1$$

and

$$\varphi_{\theta^{\mathrm{T}}}(\tilde{\beta}, \tilde{p}) = \frac{1}{\tilde{\sigma}^2} \begin{pmatrix} X^{\mathrm{T}} \\ 1_n^{\mathrm{T}} \widehat{\mathcal{E}} \end{pmatrix} \begin{pmatrix} \tilde{G}_2 X & (\tilde{G}_2 \tilde{\mathcal{E}} + \tilde{G}_1) 1_n \end{pmatrix}.$$

The elements of $X^{\mathrm{T}} \tilde{g}_1$ and of $X^{\mathrm{T}} \tilde{G}_1 1_n$ corresponding to the nuisance parameters are zero.

(c) *Normal theory linear model:*

$$\varphi(\widehat{\beta}; \widehat{p}) = -\frac{1}{\widehat{\sigma}} \begin{pmatrix} 0_p \\ n \end{pmatrix}, \quad \varphi(\tilde{\beta}; \tilde{p}) = -\frac{1}{\tilde{\sigma}} \begin{pmatrix} X^{\mathrm{T}} \\ 1_n^{\mathrm{T}} \widehat{\mathcal{E}} \end{pmatrix} \tilde{\mathcal{E}} 1_n$$

and

$$\varphi_{\theta^{\mathrm{T}}}(\tilde{\beta}, \tilde{p}) = \frac{1}{\tilde{\sigma}^2} \begin{pmatrix} X^{\mathrm{T}} \\ 1_n^{\mathrm{T}} \widehat{\mathcal{E}} \end{pmatrix} \begin{pmatrix} X & 2\tilde{\mathcal{E}} 1_n \end{pmatrix}.$$

The elements of $X^{\mathrm{T}} \tilde{\mathcal{E}} 1_n$ and of $X^{\mathrm{T}} \tilde{\mathcal{E}} 1_n$ that correspond to the nuisance parameters are zero.   (Section 8.6)

63. Verify that by adding the column $\ell_{\psi;\widehat{\theta}}$ and a suitable row vector of zeros with a single one, (8.36) gives the numerator of q_1 in (8.32).   (Section 8.5)

64. At the time of writing it is not clear what the relationship is between Skovgaard's approximation to q_1 and the Fraser–Reid–Wu (1999a) expression q_2. Some progress can be made in the nonlinear model with constant variance, where

$$\frac{\partial \phi(\theta)}{\partial \theta^{\mathrm{T}}} = W^{-3} \begin{pmatrix} W X_\beta & -2M 1_m \\ 0_{m \times p} & 1_m \end{pmatrix}$$

and

$$\frac{\partial \phi(\widehat{\theta})^{\mathrm{T}}}{\partial \theta} \Sigma(\theta) = \begin{pmatrix} X_\beta^{\mathrm{T}} N & 2 X_\beta^{\mathrm{T}} M N \\ 0_m^{\mathrm{T}} & -4 1_m^{\mathrm{T}} W^{-1} M^2 (N - I_m) + 2 1_m^{\mathrm{T}} W N \end{pmatrix},$$

where 0_m is a vector of zeros of length m. This yields

$$S(\widehat{\theta}, \tilde{\theta}) = \begin{pmatrix} \widehat{X}_\beta^{\mathrm{T}} N \tilde{W}^{-2} \tilde{X}_\beta & 2 \widehat{X}_\beta^{\mathrm{T}} N (\widehat{M} - \tilde{M}) 1_m \\ 0_p^{\mathrm{T}} & 1_m^{\mathrm{T}} (-4 \widehat{W}^{-1} \widehat{M}^2 (N - I_m) + 2 \widehat{W} N) 1_m \end{pmatrix}$$

and

$$Q(\widehat{\theta}, \tilde{\theta}) = \begin{pmatrix} \frac{1}{2}(\widehat{X}_\beta^{\mathsf{T}} N \widehat{M}(\tilde{W}^{-2} + \widehat{W}^{-2}) - \widehat{X}_\beta N \widehat{M} \tilde{W}^{-2}) 1_m \\ -\frac{1}{2} 1_m^{\mathsf{T}}(-4\widehat{W}^{-1}\widehat{M}^2(N - I_m) + 2\widehat{W}N)(\widehat{W}^{-2} - \tilde{W}^{-2}) 1_m \end{pmatrix}.$$

It may be possible to work with the expressions given at the beginning of Section 8.6.3 to compare q_1 and q_2 directly in this case. In the case of non-constant variance the two approaches seem to be quite different, as q_1 is based on the minimal sufficient statistics, whereas q_2 is based on all the observations y. Further, Q and S depend on the means μ_i, which are not explicitly needed in the expression for q_2. (Section 8.5)

65. The approximate normality of the posterior density can be established informally by writing $f(y \mid \theta) = \exp \ell(\theta \mid y)$ and expanding $\ell(\theta)$ in a Taylor series about $\widehat{\theta}$ up to quadratic terms. Adapt this technique to derive the alternative approximation

$$\theta \mid y \overset{\sim}{\sim} N\left(\widehat{\theta}_\pi, j_\pi(\widehat{\theta}_\pi)^{-1}\right),$$

where $\widehat{\theta}_\pi$ maximizes $f(y \mid \theta)\pi(\theta)$ and $j_\pi(\widehat{\theta}_\pi)$ is the curvature of $\log\{f(y \mid \theta)\pi(\theta)\}$ at the maximum.

(Section 8.7; Berger, 1985, Chapter 4)

66. Derive explicit expressions for the Bartlett correction factor and for Skovgaard's γ correction term in the special case of inference for θ in the full exponential family model $f(y; \theta) = \exp\{y^{\mathsf{T}}\theta - c(\theta) + h(y)\}$. McCullagh and Cox (1986) give a geometric interpretation of the Bartlett correction and Cordeiro (1987) gives an explicit expression for generalized linear models involving nuisance parameters. Skovgaard (2001) gives the expression for γ in this setting. (Section 8.8)

Appendix

67. The two-term Edgeworth approximation for the density of the unnormalized sum $S_n = \sum Y_i$ can be obtained from (A.2) as

$$f_{S_n}(z) \doteq \frac{1}{n^{1/2}}\phi\left(\frac{z - n\mu}{n^{1/2}\sigma}\right)\left\{1 + \frac{\rho_3}{6n^{1/2}}H_3\left(\frac{z - n\mu}{n^{1/2}\sigma}\right) + \frac{\rho_4}{24n}H_4\left(\frac{z - n\mu}{n^{1/2}\sigma}\right)\right.$$
$$\left. + \frac{\rho_3^2}{72n}H_6\left(\frac{z - n\mu}{n^{1/2}\sigma}\right)\right\}. \tag{10.10}$$

(Section A.2)

68. If we let $\phi(z; a, b^2) = b^{-1/2}\phi\{(z - a)/b\}$, and let $\kappa_r^n = \kappa_r(S_n)$ be the rth cumulant of S_n, then (10.10) is equivalent to

$$f_{S_n}(s) \doteq \phi(z; n\mu, n\sigma^2)\left\{1 + \frac{\kappa_3^n}{6}H_3(z; n\mu, n\sigma^2) + \frac{\kappa_4^n}{24}H_4(z; n\mu, n\sigma^2)\right.$$
$$\left. + \frac{\kappa_6^n}{72}H_6(z; n\mu, n\sigma^2)\right\}, \tag{10.11}$$

where

$$H_r(z; a, b^2) = (-1)^r \phi^{(r)}(z; a, b^2)/\phi(z; a, b^2).$$

When $y = (y^1, \ldots, y^d)$ is vector-valued, then the easiest version of the Edgeworth approximation is to generalize (10.11), and the result is

$$f_{S_n}(z) \doteq \phi(z; \mu, \Sigma) \left\{ 1 + \frac{1}{6} \kappa^{i,j,k} H_{ijk}(z; \mu, \Sigma) + \frac{1}{24} \kappa^{i,j,k,l} H_{ijkl}(z; \mu, \Sigma) \right.$$

$$\left. + \frac{1}{72} \kappa^{i,j,k} \kappa^{l,m,n} [10] H_{ijk}(z; \mu, \Sigma) H_{lmn}(z; \mu, \Sigma) \right\}. \tag{10.12}$$

The cumulant generating function for Y is $K_Y(t) = \log E\{\exp(t_i Y^i)\}$, where $t = (t_1, \ldots, t_d)$, and we use the summation convention, whereby summation is implied over identical upper and lower subscripts. The cumulants κ in (10.12) are the cumulants of S_n, and are arrays of the appropriate dimension: for example κ^i is the ith component of the mean vector μ, while $\kappa^{i,j}$ is the (i, j) element of the $d \times d$ covariance matrix Σ, and $\kappa^{i,j,k}$ and $\kappa^{i,j,k,l}$ are the coefficients in the multivariate Taylor series expansion of the cumulant generating function of S_n. The notation [10] in the final term is shorthand for the sum of 10 similar terms, each corresponding to one of the partitions of the six indices into two sets of three.

(Section A.2; Reid, 1988, Appendix; McCullagh, 1987, Chapter 5)

69. Derive an expression for the $O(n^{-1})$ term in the d-dimensional Laplace approximation (A.17); the notation used in McCullagh (1987, Chapter 3) may be helpful. In the scalar parameter case both the $O(n^{-1})$ and $O(n^{-2})$ terms are given in Tierney and Kadane (1986). The relationship between the $O(n^{-1})$ term and the Bartlett correction factor is explored in DiCiccio and Stern (1993). (Section A.3)

Appendix A

Some numerical techniques

A.1 Convergence of sequences

It is very useful to describe the large sample behaviour of sequences of constants or random variables using the $o(\cdot)$, $O(\cdot)$, $o_p(\cdot)$ and $O_p(\cdot)$ notation, which we now describe. We assume that $\{a_n\}$, $\{b_n\}$ and so forth are sequences of constants such as $\{1, n^{-1/2}, n^{-1}, n^{-3/2}, \dots\}$ or $\{1, n^{-1}, n^{-2}, \dots\}$, and similarly $\{X_n\}$, $\{Y_n\}$ and so forth are sequences of random variables, for $n = 1, 2, \dots$

We say $a_n = o(b_n)$ if $a_n/b_n \to 0$ as $n \to \infty$ and $a_n = O(b_n)$ if $a_n/b_n \to A$ as $n \to \infty$, where A is a finite constant.

Similarly $X_n = o_p(a_n)$ if $X_n/a_n \overset{p}{\to} 0$ as $n \to \infty$, where the notation $\overset{p}{\to}$ refers to convergence in probability: for any $\epsilon > 0$, $\lim_{n \to \infty} \Pr(|X_n/a_n| > \epsilon) = 0$. For example, if $X_n = (Y_1 + \cdots + Y_n)/n$, where Y_i are independent and each has finite mean μ, then $X_n - \mu = o_p(1)$ by the weak law of large numbers. We use the notation $\overset{d}{\to}$ to mean convergence in distribution: $X_n \overset{d}{\to} X$ if $\Pr(X_n \leq x) \to \Pr(X \leq x)$ for every x at which the cumulative distribution function of X is continuous.

The O_p notation refers to boundedness in probability, which is slightly more cumbersome to describe: $X_n = O_p(a_n)$ if for each ϵ there exists an n_0 and a finite constant A such that

$$\Pr(|X_n/a_n| > A) < \epsilon, \quad \text{for all } n > n_0.$$

If the Y_i above are identically distributed with finite variance, then the central limit theorem implies that $X_n - \mu = O_p(n^{-1/2})$, or equivalently $\sqrt{n}(X_n - \mu) = O_p(1)$.

A.2 The sample mean

Preliminaries

Assume that $Y, Y_1, \dots, Y_n$ are independent, identically distributed random variables from a distribution $F(\cdot)$ on the real line, with moment generating function $M_Y(t) = \mathrm{E}\{\exp(tY)\}$ and cumulant generating function

$$K_Y(t) = \log M_Y(t), \tag{A.1}$$

and that $K_Y(t)$ has an expansion about zero of the form

$$K_Y(t) = \kappa_1 t + \frac{1}{2!}\kappa_2 t^2 + \frac{1}{3!}\kappa_3 t^3 + \frac{1}{4!}\kappa_4 + \cdots.$$

The κ_r are the *cumulants* of Y, and can be related to the moments using (A.1); we have for example that

$$\kappa_1 = \mu = E(Y),$$
$$\kappa_2 = \sigma^2 = \text{var}(Y),$$
$$\kappa_3 = E(Y-\mu)^3,$$
$$\kappa_4 = E(Y-\mu)^4 - 3\sigma^4.$$

It is convenient to have special notation for the standardized skewness and kurtosis; we write

$$\rho_3 = \kappa_3/\sigma^{3/2}, \qquad \rho_4 = \kappa_4/\sigma^2.$$

In this section we describe approximations for the distribution of the sample mean, $\overline{Y} = n^{-1}\sum Y_i$, as $n \to \infty$. Let

$$S_n = \sum_{i=1}^{n} Y_i = n\overline{Y}, \quad S_n^* = (S_n - n\mu)/\sqrt{n}\sigma = \sqrt{n}(\overline{Y} - \mu)/\sigma$$

be the sum and the standardized sum. The limiting distribution of S_n^* is the standard normal distribution, as follows from the central limit theorem, which states that $S_n^* \xrightarrow{d} Z$, where $Z \sim N(0, 1)$, that is,

$$\Pr(S_n^* \leq z) \to \int_{-\infty}^{z} \frac{1}{(2\pi)^{1/2}} \exp(-t^2/2)\,dt$$

for all z. The first order approximation provided by this asymptotic result is that the sample mean is approximately normally distributed, with mean μ and variance σ^2/n; we write this as

$$\overline{Y} \stackrel{.}{\sim} N(\mu, \sigma^2/n).$$

Edgeworth expansion

Write $\phi(\cdot)$ and $\Phi(\cdot)$ for the density and distribution function of the standard normal, and $H_r(\cdot)$ for the rth *Hermite polynomial*, defined by

$$H_r(x) = (-1)^r \phi^{(r)}(x)/\phi(x),$$

where $\phi^{(r)}(x) = d^r\phi(x)/dx^r$.

The Edgeworth expansion for the density of S_n^* is

$$f_{S_n^*}(x) = \phi(x)\left\{1 + \frac{\rho_3}{6n^{1/2}}H_3(x) + \frac{\rho_4}{24n}H_4(x) + \frac{\rho_3^2}{72n}H_6(x)\right\} + O(n^{-3/2}), \qquad (A.2)$$

and the corresponding Edgeworth expansion for the distribution function is

$$F_{S_n^*}(x) = \Phi(x) - \phi(x)\left\{\frac{\rho_3}{6n^{1/2}}H_2(x) + \frac{\rho_4}{24n}H_3(x) + \frac{\rho_3^2}{72n}H_5(x)\right\} + O(n^{-3/2}). \qquad (A.3)$$

The two-term Edgeworth approximation to the density or distribution of S_n^* is obtained from the right-hand side of (A.2) or (A.3), ignoring the error term $O(n^{-3/2})$. Similarly the one-term Edgeworth approximation uses only the leading term and the $n^{-1/2}$ term in (A.2) or (A.3).

Saddlepoint expansion

The Edgeworth expansion is derived by expanding the moment generating function (or more generally the characteristic function) for S_n^* in a Taylor series about zero, and inverting this term by term. A different inversion of the characteristic function gives the saddlepoint expansion for the density of S_n. The result is

$$f_{S_n}(s) = \frac{1}{(2\pi)^{1/2}}\frac{1}{\{nK_Y''(\widehat{t})\}^{1/2}}\exp\{nK_Y(\widehat{t}) - \widehat{t}s\}\{1 + O(n^{-1})\}, \qquad (A.4)$$

where $\widehat{t}$ satisfies the equation $nK_Y'(\widehat{t}) = s$, assumed to exist uniquely. The $O(n^{-1})$ term in (A.4) is

$$(3\widehat{\rho}_4 - 5\widehat{\rho}_3^2)/(24n), \qquad (A.5)$$

where

$$\widehat{\rho}_j = \rho_j(\widehat{t}) = K_Y^{(j)}(\widehat{t})/\{K_Y''(\widehat{t})\}^{j/2}.$$

If the leading term of (A.4) is renormalized, it approximates the density of S_n with relative error $O(n^{-3/2})$. A simple change of variables gives a saddlepoint expansion for $\overline{Y}_n = S_n/n$.

For d-dimensional Y, the saddlepoint expansion of the density is

$$f_{S_n}(s) = \frac{1}{(2\pi)^{d/2}}\frac{1}{\{n|K_Y''(\widehat{t})|\}^{1/2}}\exp\{nK_Y(\widehat{t}) - \widehat{t}^{\mathsf{T}}s\}\{1 + O(n^{-1})\},$$

where $K_Y'(t)$ is a vector and $K_Y''(t)$ is a $d \times d$ matrix.

Approximating the cumulative distribution function

When Y is one-dimensional, an expansion for the distribution function of S_n can also be obtained by the saddlepoint technique. We will give a heuristic derivation of the approximation, starting with the renormalized version of (A.4). We have

$$
\begin{aligned}
F_{S_n}(s) &= \int_{-\infty}^{s} f_{S_n}(x)\mathrm{d}x \\
&\doteq \int_{-\infty}^{s} c\{nK_Y''(\widehat{t})\}^{-1/2}\exp\{nK_Y(\widehat{t})-\widehat{t}\,x)\}\,\mathrm{d}x,
\end{aligned}
\tag{A.6}
$$

where $\widehat{t}$ is a function of x through the saddlepoint equation $nK_Y'(\widehat{t})=x$. Let

$$
\frac{1}{2}r^2 = \widehat{t}\,x - nK_Y(\widehat{t}),
\tag{A.7}
$$

so that

$$
\begin{aligned}
r\,\mathrm{d}r &= \left\{\widehat{t}+x\frac{\mathrm{d}\widehat{t}}{\mathrm{d}x}-nK_Y'(\widehat{t})\frac{\mathrm{d}\widehat{t}}{\mathrm{d}x}\right\}\mathrm{d}x \\
&= \widehat{t}\,\mathrm{d}x.
\end{aligned}
\tag{A.8}
$$

Changing variables in (A.6) we have

$$
F_{S_n}(s) \doteq \int_{-\infty}^{r_s} c\exp(-r^2/2)\frac{r}{\widehat{t}\,\{nK_Y''(\widehat{t})\}^{1/2}}\,\mathrm{d}r,
\tag{A.9}
$$

and through (A.7) $\widehat{t}$ is a function of r. Now let $u = \widehat{t}\,\{nK_Y''(\widehat{t})\}^{1/2}$, and $c_0 = c/\sqrt{2\pi}$:

$$
\begin{aligned}
F_{S_n}(s) &\doteq \int_{-\infty}^{r_s} c_0\phi(r)(r/u)\mathrm{d}r \\
&= \int_{-\infty}^{r_s} c_0(1+r/u-1)\phi(r)\mathrm{d}r \\
&= c_0\Phi(r_s)+\int_{-\infty}^{r_s} c_0 r\phi(r)(1/u-1/r)\mathrm{d}r,
\end{aligned}
$$

and this in turn equals

$$
c_0\Phi(r_s)+c_0\phi(r_s)\left(\frac{1}{r_s}-\frac{1}{u_s}\right)+c_0\int_{-\infty}^{r_s}\phi(r)\mathrm{d}(1/r-1/u).
\tag{A.10}
$$

As s and thus $r_s \to \infty$, the left-hand side goes to 1, which shows that $c_0 = 1+O(n^{-1})$. The second term in (A.10) is at most $O(n^{-1/2})$, so the c_0 contribution to that term can also be absorbed into the remainder term. That the remainder term is $O(n^{-3/2})$ follows after showing that

$$
r = u + Au^2/n^{1/2} + Bu^3/n + O(n^{-3/2}),
\tag{A.11}
$$

and that $1/r - 1/u$ is a linear function of r to this order, so we finally have the result

$$F_{S_n}(s) = \left\{ \Phi(r_s) - \phi(r_s) \left(\frac{1}{r_s} - \frac{1}{u_s} \right) \right\} \{ 1 + O(n^{-3/2}) \}. \tag{A.12}$$

At the first step in the integration given at (A.10), we could also write

$$F_{S_n}(s) \doteq \int_{-\infty}^{r_s} \frac{c_0}{\sqrt{(2\pi)}} e^{-r^2/2 + \log(r/u)} dr$$

$$\doteq \int_{-\infty}^{r_s} \frac{c_0}{\sqrt{(2\pi)}} e^{-\{r - (1/r)\log(u/r)\}^2/2} e^{\log^2(u/r)/(2r^2)} dr,$$

leading to the equivalent approximation

$$F_{S_n}(s) = \Phi(r_s^*) \{ 1 + O(n^{-3/2}) \}, \tag{A.13}$$

where

$$r_s^* = r_s - \frac{1}{r_s} \log \frac{u_s}{r_s}.$$

These arguments assume that r_s and u_s are monotone functions of s, as they will be in regular problems, and thus the approximations obtained from (A.13) and (A.12) can equally well be regarded as approximations to the distribution of r. We normally use the approximations without the subscript s:

$$F_{S_n}(s) \doteq \Phi(r) - \phi(r) \left(\frac{1}{r} - \frac{1}{u} \right) \equiv \Phi^*(r), \tag{A.14}$$

$$\doteq \Phi(r^*), \tag{A.15}$$

with r given by (A.8), $\hat{t}$ given by $K_Y'(\hat{t}) = s$, u defined following (A.9) and

$$r^* = r - \frac{1}{r} \log \frac{u}{r}.$$

Approximation (A.14) is known as the Lugannani–Rice approximation, and (A.15) is known as Barndorff-Nielsen's r^* approximation. These figure prominently in the examples presented in the book.

There is a singularity at $s = 0$, where $r = u = 0$, in the integrand defining F_{S_n}, so a careful derivation would have $s < 0$ to obtain an approximation to the cumulative distribution function, and for $s > 0$ use the integral from s to ∞. It can be shown that the approximations (A.11) and (A.13) are continuous at 0, and their limiting value there is

$$F_{S_n}(0) \doteq \frac{1}{2} - \frac{\rho_3(0)}{6(2\pi n)^{1/2}}, \tag{A.16}$$

where $\rho_3(\hat{t})$ is defined following (A.5). In practice this means that the approximations become quite unstable near $r = s$; the numerical implications of this are discussed in Chapter 9.

This derivation only applies to scalar random variables Y. For vector Y, there is a similar approximation to the conditional distribution of the first component of S_n, given the remaining components. See the bibliographic notes.

A.3 Laplace approximation

The asymptotic technique most often used in connection with approximating posterior distributions is Laplace approximation. We assume that we have an integral expression of the form

$$\int_a^b e^{-ng(y)} dy,$$

where the function $g(\cdot)$ does not depend on n, and is assumed to have a unique minimum $\tilde{y} \in (a, b)$ where $g'(\tilde{y}) = 0$ and $g''(\tilde{y}) > 0$. By expanding the integrand about $\tilde{y}$, we have

$$\int_a^b e^{-ng(y)} dy = e^{-ng(\tilde{y})} \left(\frac{2\pi}{n}\right)^{1/2} \{g''(\tilde{y})\}^{-1/2} \left\{1 + \frac{5\tilde{\rho}_3^2 - 3\tilde{\rho}_4}{24n} + O(n^{-2})\right\},$$

where $\tilde{\rho}_3 = g'''(\tilde{y})/\{g''(\tilde{y})\}^{3/2}$ and $\tilde{\rho}_4 = g^{(4)}(\tilde{y})/\{g''(\tilde{y})\}^2$.

The argument can be generalized to integrals of the form

$$\int h(y)e^{-ng(y)} dy,$$

with the result

$$h(\tilde{y})e^{-ng(\tilde{y})} \left(\frac{2\pi}{n}\right)^{1/2} \{g''(\tilde{y})\}^{-1/2}$$

$$\times \left\{1 + \frac{5\tilde{\rho}_3^2 - 3\tilde{\rho}_4}{24n} + \frac{h''(\tilde{y})}{2g''(\tilde{y})h(\tilde{y})n} - \frac{\tilde{\rho}_3 h'(\tilde{y})/h(\tilde{y})}{2\{g''(\tilde{y})\}^{1/2}n} + O(n^{-2})\right\},$$

for which we need to assume that $h(\tilde{y}) \neq 0$.

As in the derivation of r^* in the previous section, we could also incorporate a positive function $h(\cdot)$ into the integrand, by writing

$$\int \exp[-n\{g(y) - \frac{1}{n}\log h(y)\}]dy = \int \exp\{-nq_n(y)\}dy,$$

resulting in the approximation

$$h(y^*)e^{-ng(y^*)} \left(\frac{2\pi}{n}\right)^{1/2} \{q_n''(y^*)\}^{-1/2} \left\{1 + \frac{5\rho_3^{*2} - 3\rho_4^*}{24n} + O(n^{-2})\right\},$$

where $q_n'(y^*) = 0$ and $\rho_j^* = q^{(j)}(y^*)/\{q''(y^*)\}^{j/2}$.

The d-dimensional version of the Laplace approximation is

$$\int_{R^d} h(y)e^{-ng(y)}dy = h(\tilde{y})e^{-ng(\tilde{y})} \left(\frac{2\pi}{n}\right)^{d/2} |g''(\tilde{y})|^{-1/2}\{1 + O(n^{-1})\} \qquad (A.17)$$

$$= h(y^*)e^{-ng(y^*)} \left(\frac{2\pi}{n}\right)^{d/2} |g''(y^*)|^{-1/2}\{1 + O(n^{-1})\}.$$

A.4 χ^2 approximations

For developing higher order approximations in inference for vector parameters, the following properties of the χ_d^2 distribution are useful. Denote the density of this distribution by

$$q_d(y) = \frac{1}{2^{d/2}\Gamma(d/2)} y^{d/2-1}e^{-y/2}, \quad y > 0, \ d = 1, 2, \ldots$$

Suppose that a random variable, Y_n, say, has a density given by

$$f_{Y_n}(y) = q_d(y)(1 - an^{-1}) + q_{d+2}(y)an^{-1} + O(n^{-2}).$$

Then an equivalent representation of the density is

$$f_{Y_n}(y) = q_d(y)\{1 + a(y/d - 1)n^{-1}\} + O(n^{-2}),$$

and the moment generating function is

$$M_{Y_n}(t) = (1 - 2t)^{-d/2}\{1 + 2at(1 - 2t)^{-1}n^{-1}\} + O(n^{-2}), \quad t < 1/2.$$

Furthermore, it can be shown directly that the density of $\{1 + 2a/(dn)\}Y_n$ is $q_d(\cdot)$, with error $O(n^{-2})$.

Bibliographic notes

Barndorff-Nielsen and Cox (1989) give a useful overview of a number of numerical techniques useful in asymptotic theory, and our account draws heavily from there. The multivariate Edgeworth expansion is given in Chapter 5 of McCullagh (1987), and is also summarized in Reid (1988).

The classic paper of Daniels (1954) gives a very thorough and clear account of the saddlepoint density approximation for the sample mean. The application of saddlepoint techniques to derive the tail area approximation, and the relationship of this to the derivation of Lugannani and Rice (1980), is reviewed in Daniels (1987). Skovgaard (1987) gives tail area approximations for the conditional distribution of one component of a sample mean. The asymptotic equivalence of $\Phi^*(r)$ and $\Phi(r^*)$ was established in Jensen (1992). The Laplace approximation and χ^2 approximations discussed in Sections A.3 and A.4 are taken from Barndorff-Nielsen and Cox (1989, Chapter 4).

The mathematics of the saddlepoint approximation are treated in detail in Jensen (1995), and careful derivations of several expansions are presented in Kolassa (1994). Field and Ronchetti (1990) discuss saddlepoint approximation for M-estimators. Butler (2007) gives an encyclopedic account of saddlepoint approximations with many specific applications.

References

Abe, F., Akimoto, H., Akopian, A. *et al.* (1995) Observation of top quark production in $\overline{p}$ p collisions with the collider detector at Fermilab. *Physical Review Letters* **74**, 2626–2631.

Agresti, A. and Coull, B. A. (1998) Approximate is better than "exact" for interval estimation of binomial proportions. *American Statistician* **52**, 119–126.

Agresti, A. and Min, Y. (2005) Frequentist performance of Bayesian confidence intervals for comparing proportions in 2×2 contingency tables. *Biometrics* **61**, 515–523.

Aitchison, J. and Dunsmore, I. R. (1975) *Statistical Prediction Analysis.* Cambridge: Cambridge University Press.

Andersen, P. K., Borgan, Ø., Gill, R. D. and Keiding, N. (1993) *Statistical Models Based on Counting Processes.* New York: Springer-Verlag.

Andrews, D. F. and Herzberg, A. M. (1985) *Data: A Collection of Problems from Many Fields for the Student and Research Worker.* New York: Springer-Verlag.

Andrews, D. F. and Stafford, J. E. (2000) *Symbolic Computation for Statistical Inference.* Oxford: Oxford University Press.

Annis, D. H. (2005) Rethinking the paper helicopter: Combining statistical and engineering knowledge. *The American Statistician* **59**, 320–326.

Barndorff-Nielsen, O. E. (1978) *Information and Exponential Families in Statistical Theory.* Chichester: Wiley.

Barndorff-Nielsen, O. E. (1980) Conditionality resolutions. *Biometrika* **67**, 293–310.

Barndorff-Nielsen, O. E. (1983) On a formula for the distribution of the maximum likelihood estimator. *Biometrika* **70**, 343–365.

Barndorff-Nielsen, O. E. (1986) Inference on full or partial parameters based on the standardized signed log likelihood ratio. *Biometrika* **73**, 307–322.

Barndorff-Nielsen, O. E. (1990) Approximate interval probabilities. *Journal of the Royal Statistical Society, Series B* **52**, 485–496.

Barndorff-Nielsen, O. E. and Chamberlin, S. R. (1991) An ancillary invariant modification of the signed log likelihood ratio. *Scandinavian Journal of Statistics* **18**, 341–352.

Barndorff-Nielsen, O. E. and Cox, D. R. (1979) Edgeworth and saddlepoint approximations with statistical applications (with discussion). *Journal of the Royal Statistical Society, Series B* **41**, 279–312.

Barndorff-Nielsen, O. E. and Cox, D. R. (1989) *Asymptotic Techniques for Use in Statistics.* London: Chapman & Hall.

Barndorff-Nielsen, O. E. and Cox, D. R. (1994) *Inference and Asymptotics.* London: Chapman & Hall.

Bartlett, M. S. (1937) Properties of sufficiency and statistical tests. *Proceedings of the Royal Society of London, Series A* **160**, 268–282.

Bates, D. M. and Watts, D. G. (1988) *Nonlinear Regression Analysis and Its Applications*. New York: Wiley.

Becker, R. A., Chambers, J. M. and Wilks, A. R. (1988) *The New S Language: A Programming Environment for Data Analysis and Graphics*. London: Chapman & Hall.

Belanger, B. A., Davidian, M. and Giltinan, D. M. (1996) The effect of variance function estimation on nonlinear calibration inference in immunoassay data. *Biometrics* **52**, 158–175.

Bellio, R. (2000) *Likelihood Asymptotics: Applications in Biostatistics*. Ph.D. thesis, Department of Statistics, University of Padova.

Bellio, R. (2003) Likelihood methods for controlled calibration. *Scandinavian Journal of Statistics* **30**, 339–353.

Bellio, R. and Brazzale, A. R. (1999) Higher-order asymptotics in nonlinear regression. In *Proceedings of the 14th International Workshop on Statistical Modelling*, eds. H. Friedl, A. Berghold and G. Kauermann, pp. 440–443.

Bellio, R. and Brazzale, A. R. (2001) A computer algebra package for approximate conditional inference. *Statistics and Computing* **11**, 17–24.

Bellio, R. and Brazzale, A. R. (2003) Higher-order asymptotics unleashed: Software design for nonlinear heteroscedastic models. *Journal of Computational and Graphical Statistics* **12**, 682–697.

Bellio, R. and Sartori, N. (2006) Practical use of modified maximum likelihoods for stratified data. *Biometrical Journal*, **48**, 876–886.

Bellio, R., Jensen, J. E. and Seiden, P. (2000) Applications of likelihood asymptotics for nonlinear regression in herbicide bioassays. *Biometrics* **56**, 1204–1212.

Berger, J. O. (1985) *Statistical Decision Theory and Bayesian Analysis*. Second edition. New York: Springer-Verlag.

Berkson, J. (1950) Are there two regressions? *Journal of the American Statistical Association* **45**, 164–180.

Besag, J. E. and Clifford, P. (1989) Generalized Monte Carlo significance tests. *Biometrika* **76**, 633–642.

Besag, J. E. and Clifford, P. (1991) Sequential Monte Carlo *p*-values. *Biometrika* **78**, 301–304.

Bickel, P. J. and Ghosh, J. K. (1990) A decomposition for the likelihood ratio statistic and the Bartlett correction – a Bayesian argument. *Annals of Statistics* **18**, 1070–1090.

Böhning, D., Dietz, E., Schlattman, P., Mendorça, L. and Kirchner, U. (1999), The zero-inflated Poisson model and the decayed, missing and filled teeth index in dental epidemiology. *Journal of the Royal Statistical Society, Series A* **162**, 195–209.

Booth, J. G. and Butler, R. W. (1990) Randomization distributions and saddlepoint approximations in generalized linear models. *Biometrika* **77**, 787–796.

Box, G. E. P. (1949) A general distribution theory for a class of likelihood criteria. *Biometrika* **36**, 317–346.

Box, G. E. P. (1992) Teaching engineers experimental design with a paper helicopter. *Quality Engineering* **4**, 453–459.

Box, G. E. P. and Cox, D. R. (1964) An analysis of transformations (with discussion). *Journal of the Royal Statistical Society, Series B* **26**, 211–264.

Brain, P. and Cousens, R. (1989) An equation to describe dose response where there is stimulation of growth at low doses. *Weed Research* **29**, 93–96.

Brazzale, A. R. (1999) Approximate conditional inference in logistic and loglinear models. *Journal of Computational and Graphical Statistics* **8**, 653–661.

Brazzale, A. R. (2000) *Practical Small-Sample Parametric Inference*. Ph.D. thesis, Ecole Polytechnique Fédérale de Lausanne, Switzerland.

Brazzale, A. R. (2005) hoa: An R package bundle for higher order likelihood inference. *R-News* **5/1**, 20–27.

Brown, L. D. (1986) *Fundamentals of Statistical Exponential Families, with Applications in Statistical Decision Theory*. Volume 9 of *Lecture Notes – Monograph Series*. Hayward, California: Institute of Mathematical Statistics.

Brown, L. D., Cai, T. and DasGupta, A. (2001) Interval estimation for a binomial proportion (with discussion). *Statistical Science* **16**, 101–133.

Butler, R. W. (2007) *Saddlepoint Approximations with Applications*. Cambridge: Cambridge University Press.

Butler, R. W., Huzurbazar, S. and Booth, J. G. (1992a) Saddlepoint approximations for the generalized variance and Wilks' statistic. *Biometrika* **79**, 157–169.

Butler, R. W., Huzurbazar, S. and Booth, J. G. (1992b) Saddlepoint approximations for the Bartlett–Nanda–Pillai trace statistic in multivariate analysis. *Biometrika* **79**, 705–715.

Carroll, R. J. and Ruppert, D. (1988) *Transformation and Weighting in Regression*. London: Chapman & Hall.

Castillo, J. D. and López-Ratera, A. (2006) Saddlepoint approximation in exponential models with boundary points. *Bernoulli* **12**, 491–500.

Chèvre, N., Becker-van Slooten, K., Tarradellas, J., Brazzale, A. R., Behra, R. and Guettinger, H. (2002) Effects of dinoseb on the entire life-cycle of *Daphnia magna*. Part II: Modelling of survival and proposal of an alternative to No-Observed-Effect-Concentration (NOEC). *Environmental Toxicology and Chemistry* **21**, 828–833.

Christensen, O. F., Frydenberg, M., Jensen, J. L. and Pedersen, J. G. (2007) Tests and confidence intervals for an extended variance component using the modified likelihood ratio statistic. *Scandinavian Journal of Statistics* **34**, to appear.

Cook, R. D. and Weisberg, S. (1982) *Residuals and Influence in Regression*. London: Chapman & Hall.

Cook, R. D. and Weisberg, S. (1983) Diagnostics for heteroscedasticity in regression. *Biometrika* **70**, 1–10.

Cordeiro, G. M. (1987) On the corrections to the likelihood ratio statistics. *Biometrika* **74**, 265–274.

Cordeiro, G. M. and Aubin, E. C. Q. (2003) Bartlett adjustments for two-parameter exponential family models. *Journal of Statistical Computation and Simulation* **73**, 807–817.

Cordeiro, G. M. and Cribari-Neto, F. (1996) On Bartlett and Bartlett-type corrections. *Econometric Reviews* **15**, 339–367.

Cox, D. R. (1970) *Analysis of Binary Data*. London: Chapman & Hall.

Cox, D. R. (1980) Local ancillarity. *Biometrika* **67**, 279–286.

Cox, D. R. and Hinkley, D. V. (1974) *Theoretical Statistics*. London: Chapman & Hall.

Cox, D. R. and Oakes, D. (1984) *Analysis of Survival Data*. London: Chapman & Hall.

Cox, D. R. and Snell, E. J. (1968) A general definition of residuals (with discussion). *Journal of the Royal Statistical Society, Series B* **30**, 248–275.

Cox, D. R. and Snell, E. J. (1981) *Applied Statistics: Principles and Examples*. London: Chapman & Hall.

Cox, D. R. and Solomon, P. J. (2003) *Components of Variance*. London: Chapman & Hall.

Cox, D. R. and Wermuth, N. (1996) *Multivariate Dependencies: Models, Analysis and Interpretation*. London: Chapman & Hall.

Cribari-Neto, F. and Ferrari, S. L. P. (1995) Second order asymptotics for score tests in generalised linear models. *Biometrika* **82**, 426–432.

Cribari-Neto, F., Ferrari, S. L. P. and Cordeiro, G. M. (2000) Improved heteroscedasticity-consistent covariance matrix estimators. *Biometrika* **87**, 907–918.

Currie, I. D. (1995) Maximum likelihood estimation and *Mathematica*. *Applied Statistics* **44**, 379–394.

Daniels, H. E. (1954) Saddlepoint approximations in statistics. *Annals of Mathematical Statistics* **25**, 631–650.

Daniels, H. E. (1958) Discussion of "The regression analysis of binary sequences", by D. R. Cox. *Journal of the Royal Statistical Society, Series B* **20**, 236–238.

Daniels, H. E. (1987) Tail probability approximations. *International Statistical Review* **54**, 37–48.

Davison, A. C. (1986) Approximate predictive likelihood. *Biometrika* **73**, 323–332.

Davison, A. C. (1988) Approximate conditional inference in generalized linear models. *Journal of the Royal Statistical Society, Series B* **50**, 445–461.

Davison, A. C. (2003) *Statistical Models*. Cambridge: Cambridge University Press.

Davison, A. C. and Hinkley, D. V. (1988) Saddlepoint approximations in resampling methods. *Biometrika* **75**, 417–431.

Davison, A. C. and Hinkley, D. V. (1997) *Bootstrap Methods and Their Application*. Cambridge: Cambridge University Press.

Davison, A. C. and Wang, S. (2002) Saddlepoint approximations as smoothers. *Biometrika* **89**, 933–938.

Davison, A. C., Fraser, D. A. S. and Reid, N. (2006) Improved likelihood inference for discrete data. *Journal of the Royal Statistical Society, Series B* **68**, 495–508.

Deng, D. and Paul, S. R. (2005) Score tests for zero-inflation and over-dispersion in generalized linear models. *Statistica Sinica* **15**, 257–276.

DiCiccio, T. J. and Efron, B. (1996) Bootstrap confidence intervals (with discussion). *Statistical Science* **11**, 189–228.

DiCiccio, T. J. and Martin, M. A. (1991) Approximations of marginal tail probabilities and inference for scalar parameters. *Biometrika* **78**, 891–902.

DiCiccio, T. J. and Martin, M. A. (1993) Simple modifications for signed roots of likelihood ratio statistics. *Journal of the Royal Statistical Society, Series B* **55**, 305–316.

DiCiccio, T. J., Field, C. A. and Fraser, D. A. S. (1990) Approximations of marginal tail probabilities and inference for scalar parameters. *Biometrika* **77**, 77–95.

DiCiccio, T. J. and Stern, S. E. (1993) On Bartlett adjustments for approximate Bayesian inference. *Biometrika* **80**, 731–740.

DiCiccio, T. J., Martin, M. A. and Stern, S. E. (2001) Simple and accurate one-sided inference from signed roots of likelihood ratios. *Canadian Journal of Statistics* **29**, 67–76.

Durbin, J. (1980a) Approximations for densities of sufficient estimators. *Biometrika* **67**, 311–333.

Durbin, J. (1980b) The approximate distribution of partial serial correlation coefficients calculated from residuals from regression on Fourier series. *Biometrika* **67**, 335–349.

Efron, B. (1993) Bayes and likelihood calculations from confidence intervals. *Biometrika* **80**, 3–26.

Efron, B. and Hinkley, D. V. (1978) Assessing the accuracy of the maximum likelihood estimator: Observed versus expected Fisher information (with discussion). *Biometrika* **65**, 457–487.

Efron, B. and Tibshirani, R. J. (1993) *An Introduction to the Bootstrap*. New York: Chapman & Hall.

Eno, D. R. (1999) Noninformative prior Bayesian analysis for statistical calibration problems. Ph.D. Thesis, Virginia Polytechnic University.

Evans, K., Tyrer, P., Catalan, J. *et al.* (1999) Manual-assisted-cognitive-behaviour therapy in the treatment of recurrent deliberate self harm: A randomised controlled trial. *Psychological Medicine* **29**, 19–25.

Feigl, P. and Zelen, M. (1965) Estimation of exponential survival probabilities with concomitant information. *Biometrics* **21**, 826–838.

Feldman, G. J. and Cousins, R. D. (1998) Unified approach to the classical statistical analysis of small signals. *Physical Review, D* **57**, 3873–3889.

Ferrari, S. L. P., Cordeiro, G. M. and Cribari-Neto, F. (2001) Higher-order asymptotic refinements for score tests in proper dispersion models. *Journal of Statistical Planning and Inference* **97**, 177–190.

Ferrari, S. L. P., Lucambio, F. and Cribari-Neto, F. (2005) Improved profile likelihood inference. *Journal of Statistical Planning and Inference* **134**, 373–391.

Field, C. A. and Ronchetti, E. M. (1990) *Small Sample Asymptotics*. Volume 13 of *Lecture Notes – Monograph Series*. Hayward, California: Institute of Mathematical Statistics.

Field, C. A. and Welsh, A. H. (1998) Robust confidence intervals for regression parameters. *Australian & New Zealand Journal of Statistics* **40**, 53–64.

Firth, D. (1993) Bias reduction of maximum likelihood estimates. *Biometrika* **80**, 27–38.

Fisher, R. A. (1934) Two new properties of mathematical likelihood. *Proceedings of the Royal Society of London, Series A* **144**, 285–307.

Fleming, T. R. and Harrington, D. P. (1991) *Counting Processes and Survival Analysis*. New York: John Wiley.

Forster, J. J., McDonald, J. W. and Smith, P. W. F. (1996) Monte Carlo exact conditional tests for log-linear and logistic models. *Journal of the Royal Statistical Society, Series B* **58**, 445–453.

Fraser, D. A. S. (1979) *Inference and Linear Models*. New York: McGraw Hill.

Fraser, D. A. S. (1990) Tail probabilities from observed likelihoods. *Biometrika* **77**, 65–76.

Fraser, D. A. S. (1991) Statistical inference: Likelihood to significance. *Journal of the American Statistical Association* **86**, 258–265.

Fraser, D. A. S. (2003) Likelihood for component parameters. *Biometrika* **90**, 327–339.

Fraser, D. A. S. and Reid, N. (1993) Third order asymptotic models: Likelihood functions leading to accurate approximations to distribution functions. *Statistica Sinica* **3**, 67–82.

Fraser, D. A. S. and Reid, N. (1995) Ancillaries and third order significance. *Utilitas Mathematica* **47**, 33–53.

Fraser, D. A. S. and Reid, N. (2001) Ancillary information for statistical inference. In *Empirical Bayes and Likelihood Inference*, eds E. Ahmed and N. Reid, pp. 185–210. New York: Springer-Verlag.

Fraser, D. A. S., Reid, N. and Wong, A. C. M. (1991) Exponential linear models: A two-pass procedure for saddlepoint approximation. *Journal of the Royal Statistical Society, Series B* **53**, 483–492.

Fraser, D. A. S., Reid, N. and Wong, A. C. M. (2004) Inference for bounded parameters: A different perspective. *Physical Review, D* **69**, 033002.

Fraser, D. A. S., Reid, N. and Wu, J. (1999a) A simple general formula for tail probabilities for frequentist and Bayesian inference. *Biometrika* **86**, 249–264.

Fraser, D. A. S., Wong, A. C. M. and Wu, J. (1999b) Regression analysis, nonlinear or nonnormal: Simple and accurate *p* values from likelihood analysis. *Journal of the American Statistical Association* **94**, 1286–1295.

Frome, E. L. (1983) The analysis of rates using Poisson regression models. *Biometrics* **39**, 665–674.

Frydenberg, M. and Jensen, J. L. (1989) Is the improved likelihood ratio statistic really improved in the discrete case? *Biometrika* **76**, 655–661.

Guolo, A. and Brazzale, A. R. (2005) Advances in small sample parametric inference for linear mixed effects models. In *Proceedings of the S.Co. 2005 Workshop "Modelli Complessi e Metodi Computazionali Intensivi per la Stima e la Previsione"*, pp. 103–108. Brixen, September 15–17. Paclova: Cleup Editrice.

Guolo, A., Brazzale, A. R. and Salvan, A. (2006) Improved inference on a scalar fixed effect of interest in nonlinear mixed-effects models. *Computational Statistics and Data Analysis*, **51**, 1602–1613.

Hampel, F. R., Ronchetti, E. M., Rousseeuw, P. J. and Stahel, W. A. (1986) *Robust Statistics: The Approach Based on Influence Functions*. New York: Wiley.

Hinkley, D. V. (1980) Likelihood as approximate pivotal distribution. *Biometrika* **67**, 287–292.

Hirose, H. (1993) Estimation of threshold stress in accelerated life-testing. *IEEE Transactions on Reliability* **42**, 650–657.

Hosmer, D. W. and Lemeshow, S. (1989) *Applied Logistic Regression*. Wiley: New York.

Hurvich, C. M. and Tsai, C.-L. (1989) Regression and time series model selection in small samples. *Biometrika* **76**, 297–307.

Hurvich, C. M. and Tsai, C.-L. (1991) Bias of the corrected AIC criterion for underfitted regression and time series models. *Biometrika* **78**, 499–509.

Jensen, J. L. (1992) The modified signed log likelihood statistic and saddlepoint approximations. *Biometrika* **79**, 693–704.

Jensen, J. L. (1995) *Saddlepoint Approximations*. Oxford: Oxford University Press.

Jensen, J. L. (1997) A simple derivation of r^* for curved exponential families. *Scandinavian Journal of Statistics* **24**, 33–46.

Jiang, L. and Wong, A. C. M. (2007) A note on inference for $\Pr(X < Y)$ for right truncated exponentially distributed data. *Statistical Papers* **48**, to appear.

Jørgensen, B. (1993) A review of conditional inference: Is there a universal definition of nonformation? *Bulletin of the International Statistical Institute* **55**, 323–340.

Kabaila, P. (1993) A method for the computer calculation of Edgeworth expansions for smooth function models. *Journal of Computational and Graphical Statistics* **2**, 199–207.

Kendall, W. S. (1992) Computer algebra and yoke geometry I: When is an expression a tensor? Technical Report 238, Department of Statistics, Warwick University.

Kendall, W. S. (1998) Computer algebra. In *Encyclopedia of Biostatistics*, eds P. Armitage and T. Colton, pp. 839–845. Chichester: Wiley.

Klein, J. P. and Moeschberger, M. L. (1997) *Survival Analysis: Techniques for Censored and Truncated Data*. New York: Springer-Verlag.

Knight, K. (2000) *Mathematical Statistics*. London: Chapman & Hall.

Kolassa, J. E. (1994) *Series Approximation Methods in Statistics*. New York: Springer-Verlag.

Lange, K. L., Little, R. J. A. and Taylor, J. M. G. (1989) Robust statistical modelling using the *t* distribution. *Journal of the American Statistical Association* **84**, 881–896.

Lawless, J. F. (2003) *Statistical Models and Methods for Lifetime Data*. Second edition. New York: Wiley.

Lawley, D. N. (1956) A general method for approximating the distribution of likelihood ratio criteria. *Biometrika* **43**, 295–303.

Lee, S. M. S. and Young, G. A. (2005) Parametric bootstrapping with nuisance parameters. *Statistics and Probability Letters* **71**, 143–153.

Lehmann, E. L. (1975) *Nonparametrics: Statistical Methods Based on Ranks*. San Francisco: Holden-Day.

Lehmann, E. L. (1986) *Testing Statistical Hypotheses*. Second edition. New York: Wiley.

Lipsitz, S. R., Dear, K. G. B., Laird, N. M. and Molenberghs, G. (1998) Tests for homogeneity of the risk difference when data are sparse. *Biometrics* **54**, 148–160.

Little, R. J. A. (2006) Calibrated Bayes: A Bayes/Frequentist roadmap. *The American Statistician* **60**, 213–223.

Lugannani, R. and Rice, S. (1980) Saddlepoint approximation for the distribution of the sum of independent random variables. *Advances in Applied Probability* **12**, 475–490.

Lyons, B. and Peters, D. (2000) Applying Skovgaard's modified directed likelihood statistic to mixed linear models. *Journal of Statistical Computation and Simulation* **65**, 225–242.

Mandelkern, M. (2002) Setting confidence intervals for bounded parameters (with discussion). *Statistical Science* **17**, 149–172.

McCullagh, P. (1987) *Tensor Methods in Statistics*. London: Chapman & Hall.

McCullagh, P. (2002) What is a statistical model? (with discussion). *Annals of Statistics* **30**, 1225–1310.

McCullagh, P. and Cox, D. (1986) Invariants and likelihood ratio statistics. *Annals of Statistics* **14**, 1419–1430.

McCullagh, P. and Nelder, J. A. (1989) *Generalized Linear Models*. Second edition. London: Chapman & Hall.

McQuarrie, A. D. R. and Tsai, C.-L. (1998) *Regression and Time Series Model Selection*. Singapore: World Scientific.

Neumann, P., O'Shaughnessy, M., Remis, R., Tsoukas, C., Lepine, D. and Davis, M. (1990) Laboratory evidence of active HIV-1 infection in Canadians with hemophilia associated with administration of heat-treated factor VIII. *Journal of Acquired Immune Deficiency Syndromes* **3**, 278–281.

Pace, L. and Salvan, A. (1997) *Principles of Statistical Inference: From a Neo-Fisherian Perspective*. Singapore: World Scientific.

Pace, L. and Salvan, A. (1999) Point estimation based on confidence intervals: Exponential families. *Journal of Statistical Computation and Simulation* **64**, 1–21.

Peers, H. W. (1965) On confidence points and Bayesian probability points in the case of several parameters. *Journal of the Royal Statistical Society, Series B* **27**, 9–16.

Peterson, D. R., Chinn, N. M. and Fisher, L. D. (1980) The sudden infant death syndrome: repetitions in families. *Journal of Pediatrics* **97**, 265–267.

Petrie, A. (1978) *Individuality in Pain and Suffering*. Second edition. Chicago: University of Chicago Press.

Pierce, D. A. and Peters, D. (1992) Practical use of higher order asymptotics for multiparameter exponential families (with discussion). *Journal of the Royal Statistical Society, Series B* **54**, 701–737.

Proschan, F. (1963) Theoretical explanation of observed decreasing failure rate. *Technometrics* **5**, 375–383.

R Development Core Team (2005) *R: A Language and Environment for Statistical Computing.* R Foundation for Statistical Computing, Vienna, Austria.

Rao, C. R. (1973) *Linear Statistical Inference and its Applications.* Second edition. New York: Wiley.

Rawlings, J. O. (1988) *Applied Regression Analysis: A Research Tool.* Pacific Grove, California: Wadsworth & Brooks/Cole.

Redelmeier, D. A. and Tibshirani, R. J. (1997a) Association between cellular telephone calls and motor vehicle collisions. *New England Journal of Medicine* **336**, 453–458.

Redelmeier, D. A. and Tibshirani, R. J. (1997b) Is using a cell-phone like driving drunk? *Chance* **10**, 5–9.

Reid, N. (1988) Saddlepoint methods and statistical inference (with discussion). *Statistical Science* **3**, 213–238.

Reid, N. (1996) Higher order asymptotics and likelihood: a review and annotated bibliography. *Canadian Journal of Statistics* **24**, 141–166.

Reid, N. (2003) Asymptotics and the theory of inference. *Annals of Statistics* **31**, 1695–1731.

Reid, N. (2005) Theoretical statistics and asymptotics. In *Celebrating Statistics: Papers in Honour of D. R. Cox*, eds A. C. Davison, Y. Dodge and N. Wermuth, pp. 165–177. Oxford: Oxford University Press.

Ridout, M. S. (1994) A comparison of confidence interval methods for dilution series experiments. *Biometrics* **50**, 289–296.

Ridout, M. S., Faddy, M. J. and Solomon, M. G. (2006) Modelling the effects of repellent chemicals on foraging bees. *Applied Statistics* **55**, 63–75.

Ronchetti, E. M. and Ventura, L. (2001) Between stability and higher-order asymptotics. *Statistics and Computing* **11**, 67–73.

Rose, C. and Smith, M. D. (2002) *Mathematical Statistics with Mathematica.* New York: Springer-Verlag.

Rudemo, M., Ruppert, D. and Streibig, J. C. (1989) Random-effect models in nonlinear regression with applications to bioassay. *Biometrics* **45**, 349–362.

Sartori, N. (2003a) Modified profile likelihoods in models with stratum nuisance parameters. *Biometrika* **90**, 533–549.

Sartori, N. (2003b) A note on likelihood asymptotics for normal linear regression. *Annals of the Institute of Statistical Mathematics* **55**, 187–195.

Seber, G. A. F. and Wild, C. J. (1989) *Nonlinear Regression.* New York: Wiley.

Seiden, P., Kappel, D. and Streibig, J. C. (1998) Response of *Brassica napus L.* tissue culture to metsulfuron methyl and chlorsulfuron. *Weed Research* **38**, 221–228.

Self, S. G. and Liang, K. Y. (1987) Asymptotic properties of maximum likelihood estimators and likelihood ratio tests under nonstandard conditions. *Journal of the American Statistical Association* **82**, 605–610.

Sen, A. and Srivastava, M. (1990) *Regression Analysis: Theory, Methods and Applications.* New York: Springer-Verlag.

Severini, T. A. (1999) An empirical adjustment to the likelihood ratio statistic. *Biometrika* **86**, 235–248.

Severini, T. A. (2000a) *Likelihood Methods in Statistics.* Oxford: Oxford University Press.

Severini, T. A. (2000b) The likelihood ratio approximation to the conditional distribution of the maximum likelihood estimator in the discrete case. *Biometrika* **87**, 939–945.

Shenton, L. R. and Bowman, K. O. (1977) *Maximum Likelihood Estimation in Small Samples.* London: Griffin.

Simonoff, J. S. and Tsai, C.-L. (1994) Use of modified profile likelihood for improved tests of constancy of variance in regression. *Applied Statistics* **43**, 357–370.

Skovgaard, I. M. (1985) Large deviation approximations for maximum likelihood estimators. *Probability and Mathematical Statistics* **6**, 89–107.

Skovgaard, I. M. (1987) Saddlepoint expansions for conditional distributions. *Journal of Applied Probability* **24**, 875–887.

Skovgaard, I. M. (1990) On the density of minimum contrast estimators. *Annals of Statistics* **18**, 779–789.

Skovgaard, I. M. (1996) An explicit large-deviation approximation to one-parameter tests. *Bernoulli* **2**, 145–165.

Skovgaard, I. M. (2001) Likelihood asymptotics. *Scandinavian Journal of Statistics* **28**, 3–32.

Smith, P. W. F., Forster, J. J. and McDonald, J. W. (1996) Monte Carlo exact tests for square contingency tables. *Journal of the Royal Statistical Society, Series A* **159**, 309–321.

Snell, E. J. (1987) *Applied Statistics: A Handbook of BMDP Analyses.* London: Chapman & Hall.

Stallard, N., Gravenor, M. B. and Curnow, R. N. (2006) Estimating numbers of infectious units from serial dilution assays. *Applied Statistics* **55**, 15–30.

Stern, S. E. (1997) A second-order adjustment to the profile likelihood in the case of a multidimensional parameter. *Journal of the Royal Statistical Society, Series B* **59**, 653–665.

Svensson, A. (1981) On a goodness-of-fit test for multiplicative Poisson models. *Annals of Statistics* **9**, 697–704.

Sweeting, T. J. (2005) On the implementation of local probability matching priors for interest parameters. *Biometrika* **92**, 47–57.

Tempany, C. M. Zhou, X., Zerhouni, E. A. *et al.* (1994) Staging of prostate cancer: results of Radiology Diagnostic Oncology Group project comparison of three MR imaging techniques. *Radiology* **192**, 47–54.

Therneau, T. M. and Grambsch, P. M. (2000) *Modeling Survival Data: Extending the Cox Model.* New York: Springer-Verlag.

Tian, L. and Wilding, G. E. (2005) Confidence intervals of the ratio of means of two independent inverse Gaussian distributions. *Journal of Statistical Planning and Inference* **133**, 381–386.

Tibshirani, R. J. (1989) Noninformative priors for one parameter of many. *Biometrika* **76**, 604–608.

Tibshirani, R. J. and Redelmeier, D. A. (1997) Cellular telephones and motor-vehicle collisions: some variations on matched-pairs analysis. *Canadian Journal of Statistics* **25**, 581–591.

Tierney, L. and Kadane, J. B. (1986) Accurate approximations for posterior moments and marginal densities. *Journal of the American Statistical Association* **81**, 82–86.

Uusipaikka, E. (1998) ELiISA: Likelihood based statistical inference using *mathematica*. Poster presented at the 1998 Joint Statistical Meetings. Code available from the author.

Wald, A. (1949) Note on the consistency of the maximum likelihood estimate. *Annals of Mathematical Statistics* **20**, 595–601.

Wong, A. C. M. and Wu, J. (2000) Practical small-sample asymptotics for distributions used in life-data analysis. *Technometrics* **42**, 149–156.

Wong, A. C. M. and Wu, J. (2001) Approximate inference for the factor loading of a simple factor analysis model. *Scandinavian Journal of Statistics* **28**, 407–414.

Wong, A. C. M. and Wu, J. (2002) Small sample asymptotic inference for the coefficient of variation: normal and non-normal models. *Journal of Statistical Planning and Inference* **104**, 73–82.

Wu, J., Jiang, G., Wong, A. C. M. and Sun, X. (2002) Likelihood analysis for the ratio of means of two independent log-normal distributions. *Biometrics* **58**, 463–496.

Wu, J., Wong, A. C. M. and Jiang, G. (2003) Likelihood-based confidence intervals for a log-normal mean. *Statistics in Medicine* **22**, 1849–1860.

Wu, J., Wong, A. C. M. and Ng, K. (2005) Likelihood-based confidence interval for the ratio of scale parameters of two independent Weibull distributions. *Journal of Statistical Planning and Inference* **135**, 487–497.

Yates, F. (1984) Tests of significance for 2×2 contingency tables (with discussion). *Journal of the Royal Statistical Society, Series A* **147**, 426–463.

Yi, G. Y., Wu, J. and Liu, Y. (2002) Implementation of higher-order asymptotics to S-Plus. *Computational Statistics and Data Analysis* **40**, 775–800.

Yubin, T., Guoying, L. and Jie, Y. (2005) Saddlepoint approach to inference for response probabilities under the logistic response model. *Journal of Statistical Planning and Inference* **133**, 405–416.

Zhou, X.-H., Tsao, M. and Qin, G. (2004) New intervals for the difference between two independent binomial proportions. *Journal of Statistical Planning and Inference* **123**, 97–115.

Example index

Name index

Index

Printed in the United States
by Baker & Taylor Publisher Services